Mankind and the oceans

Mankind and the oceans

Edited by Nobuyuki Miyazaki, Zafar Adeel, and
Kouichi Ohwada

 **United Nations
University Press**

TOKYO · NEW YORK · PARIS

The views expressed in this publication are those of the authors and do not necessarily reflect the views of the United Nations University.

United Nations University Press
United Nations University, 53-70, Jingumae 5-chome,
Shibuya-ku, Tokyo, 150-8925, Japan
Tel: +81-3-3499-2811 Fax: +81-3-3406-7345
E-mail: sales@hq.unu.edu general enquiries: press@hq.unu.edu
www.unu.edu

United Nations University Office at the United Nations, New York
2 United Nations Plaza, Room DC2-2062, New York, NY 10017, USA
Tel: +1-212-963-6387 Fax: +1-212-371-9454
E-mail: unuona@ony.unu.edu

United Nations University Press is the publishing division of the United Nations University.

Cover design by Rebecca S. Neimark, Twenty-Six Letters
Cover Photograph by Digital Vision/Getty Images

Printed in the United States of America

UNUP-1057
ISBN 92-808-1057-X

Library of Congress Cataloging-in-Publication Data

Mankind and the oceans / edited by Nobuyuki Miyazaki, Zafar Adeel, and Kouichi Ohwada.
    p. cm.
  Includes index.
  ISBN 92-808-1057-X (pbk.)
    1. Marine pollution—Prevention. 2. Marine ecology. 3. Oceanography—Research. 4. Environmental protection—International cooperation.   I. Miyazaki, Nobuyuki. II. Adeel, Zafar. III. Ohwada, Kouichi.

GC1085.M25 2004
333.91′64—dc22
                                                                2004021286

Contents

List of tables and figures WITHDRAWN

Tables

Figures

Preface

Since the Industrial Revolution in the eighteenth century, human society has evolved numerous technologies for a modern lifestyle, and apparently changed human environment to a more convenient and more efficient form. However, in recent years we have come face to face with severe global environmental problems like increasing marine pollution, the greenhouse effect, acid rain, destruction of the ozone layer, increasing desertification, etc. We need to reconsider our lifestyles and approaches comprehensively in the context of these global environmental issues.

The United Nations Conference on Environment and Development, held at Rio de Janeiro, Brazil in 1992, created a new awareness of the fundamental problems of sustainable development. The Rio Summit agreed on a broad programme of action leading into the twenty-first century – Agenda 21. Through the joint efforts of governments, non-governmental organizations, and the scientific community, we have begun to realize that global environmental problems in the long term may threaten human survival itself. We need to tackle them urgently, and the responses have to be on a global scale – this is an essential challenge for the UN system in the new millennium.

Human life has a long history of close relationship with the oceans. Oceans are the most important environmental zone for maintaining normal global environmental conditions, because they occupy about 70 per cent of the global surface. It is quite appropriate that the United Nations declared 1998 as the International Year of the Ocean. Celebrating the year had a special meaning in Japan, where a national holiday for "the Marine Day" was already established on 20 July in 1996 to enhance public awareness on oceans. This holiday marked the coming into force of the United Nations Convention on the Law of the Sea (UNCLOS) in 1996.

In this context, an international conference on "Man and the Ocean" was organized at Tokyo, Morioka, and Kamaishi in Japan from 28 October to 2 November 1998 with the cooperation of the UNU (United Nations University), ORI (Ocean Research Institute of the University of Tokyo), and IPG (Iwate Prefectural Government). Additionally, UNESCO (United Nations Educational, Scientific, and Cultural Organization) and UNEP (United Nations Environment Programme) also supported the conference.

About 50 distinguished scientists from 16 countries participated in the conference, made presentations on the most current topics and key issues pertaining to the oceanic and coastal environment, and exchanged useful information on developing a good relationship between man and the ocean. The conference had 35 distinguished scientific presentations, and was divided into two main symposia: activities related to the oceanic and coastal environments; and marine pollution and biodiversity. During the conference we shared our views on the current worldwide and local problems of marine pollution and biodiversity as well as impacts on the marine ecosystems. The practical social-political strategies for resolving the issues of marine environment, marine pollution, and biodiversity for the coming century were also discussed.

These discussions amongst the participants led to a consensus that we should establish a new international research programme, to be launched in the spirit of cooperation observed during the conference. These recommendations materialized into action when we started a four-year international cooperative research project on the marine environment on 20 August 1999, financially supported by the IPG, based on an agreement among the UNU, ORI, and IPG. This project is composed of three scientific research areas – coastal marine ecosystems; nutrient cycles in the coastal waters; and marine pollution on local and global scales – and two networking initiatives – network development between scientists and the general public; and networking for public education and policy-relevant output. A programme website was established (www.pref.iwate.jp/~hp020901/) on 1 March 2000 and started to disseminate the project results.

In order to disseminate the research work presented at the "Man and the Ocean" conference and to encourage initiatives for resolving current marine environmental crises, it was decided to publish the papers presented at the conference as a book. The concept of this book, entitled *Mankind and the Oceans*, is that sustainable human development requires healthy functioning ecosystems. The principal objective of this book is to explore the relationship between human life and the ocean from aspects of marine environment, marine pollution, marine biodiversity, and the desirable management approaches. This volume comprises 12 chapters, which focus on three thematic areas: human activities related to marine life and management, case studies of marine pollution, and research on marine biodiversity and environment. It is intended as a reference book for undergraduate and graduate students, as well as the general public.

The three segments of the book highlight excellent case studies on marine environment and biodiversity conservation, as well as approaches for resolving these problems. We know that these marine environmental problems are intertwined in a complex way. However, we feel strongly that by striving together, human ingenuity can find innovative and intelligent solutions to these problems and that now is the right time to put these ideas into action. We hope this work is a first meaningful step along the path for a new wave of research leading to reconsideration of "Man and the Ocean".

We thank all of the participants and supporting staff at the "Man and the Ocean" conference, and the UNU, ORI, IPG, UNESCO, and UNEP for their cooperation in making this book. We sincerely appreciate Rector J. A. van Ginkel of the UNU, Governor Hiroya Masuda of the IPG, and former Director Keisuke Taira of the ORI for their support of publication of this book. Finally, we would like to thank Fumiko Sasaki for her valuable assistance in formatting of manuscripts, and Gareth Johnston, Scott McQuade, and Yoko Kojima of UNU Press for their continuous help and encouragement during this book's production.

Nobuyuki Miyazaki
Zafar Adeel
Kouichi Ohwada

1

Overview of the global marine and coastal challenges

Zafar Adeel and Nobuyuki Miyazaki

Introduction

The earth is frequently referred to as the blue planet, an indirect reference to how the globe appears from space; the blue colour obviously imparted by the oceans and seas circumscribing the world. In addition to imparting the blue colour, oceans govern a vast majority of the processes on the earth, including the hydrological cycle, the solar energy balance, the global nutrient cycle, the biological food chain, and the global and regional climate patterns. It is estimated that there is about 1,386 million km^3 of water on earth (Korzoun, 1978; Shiklomanov, 2000); about 97.5 per cent of this amount is seawater and only 2.5 per cent is freshwater. This huge volume of water, when linked to the very large input of solar energy (at the rate of 240 W/m^2), becomes the "boiler room" for the global climate (Voituriez and Jacques, 2000). Ocean currents are also generated as a result of the uneven distribution of solar energy. These processes, when combined, result in the distribution of water and nutrients that in turn drives the global biological cycles.

These natural physical, chemical, and biological ocean-driven processes have a great impact on human society and the various ecosystems. In addition, a vast majority of human activities are directly or indirectly dependent on oceans, including shipping and transportation, fisheries and food supply, recreation and tourism, and offshore exploration for minerals and petroleum. Therefore, understanding the oceanic processes as well as the living resources contained therein is critical to our survival.

1

It is ironic that by the time the United Nations Convention on the Law of the Sea (UNCLOS) came into force on 16 November 1994, marine and coastal areas were under threat from various anthropogenic factors (Borgese, 1998). This volume highlights some of these factors, focusing on regional and national case studies. The emphasis is also placed on finding approaches that can help remedy the impacts of anthropogenic factors.

Outline of the book

This volume is divided into three broad sections. Part 1 focuses on human activities related to the marine and coastal environment. It provides insights into our fundamental understanding of the relationship between human society and the ocean, and the drivers behind the "Mankind and the Oceans" concept. Part 2 of the book addresses marine environmental problems in various areas in the world, and reviews the current knowledge of pollution and biological impacts by hazardous chemicals like organochlorine compounds, organotins, and heavy metals. These specific case studies help us visualize the overall impact on various species and ecosystems. Part 3 addresses the current status of biodiversity and environmental problems in the Black Sea and the south-western Atlantic Ocean. The current problems of biodiversity and environment in typical enclosed- and open-sea areas are discussed.

In Part 1, Doumenge highlights the close relationship between marine pollution and high coastal human densities and related economic activities, while discussing some anthropological problems like overfishing and tourism pressures. He insists that establishing a global policy framework is critical in order to manage the sea resources efficiently. To observe such a policy framework on a regional scale, Handa summarizes the international activities of marine scientists undertaken with Japanese initiatives. A particular focus of these initiatives is on understanding the global biogeochemical cycle of bio-elements, mainly focusing on carbon and the likely changes. A more global view is presented by Uitto and Adeel, who discuss the importance of management for global marine resources through a coherent effort by international organizations. They introduce the activities of the United Nations University (UNU) on marine environment and demonstrate the utility of international networks to achieve integrated coastal resource management.

More localized examples of policy and management approaches are also included in Part 1. Okaichi and Yamada explain the environmental conditions for the Seto Inland Sea, Japan, where the coastal area has been remarkably polluted by industrial activity since 1955. They provide an overview of the environmental management approaches for this enclosed sea, while highlighting the cooperation of domestic and international bodies.

In Part 2, Zhou presents the recent environmental problems of pollutants and excess nutrients, such as inorganic nitrogen, inorganic phosphorous, and oil, in the coastal waters of China. He also describes the legal provisions and measures to prevent, reduce, and control pollution of the marine environment from land-based sources. Prudente *et al.* present monitoring data of organotin and organochlorine compounds in green mussels collected in the coastal areas of Thailand, the Philippines, and India during the period 1994–1997 – this effort is part of the International Mussel Watch Programme. They suggest that organotin contamination levels in Asian developing countries are lower than those in more developed areas and that coastal areas in the Asia-Pacific region are still being polluted by organochlorine compounds. Nakata *et al.* report high accumulation of organochlorine compounds in Baikal seals, estimate the transfer rate of the chemicals from mother to pup, and suggest the species has a high risk of toxic impact by coplanar polychlorinated biphenyls (PCBs). Similarly, O'Shea reviews the current state of knowledge on contaminants (organochlorine pesticides, metals, and butyltins) accumulated in about 80 species of marine mammals that inhabit the Pacific Ocean. He stresses the importance of improved coordination and planning of contaminant studies for understanding future trends and impacts of contaminants on marine mammals of the Pacific Ocean. Ross reports several virus-associated mass mortalities of marine mammals, and reviews studies on possible links between mass mortalities and chemical contaminants, immunotoxicity, and these outbreaks of disease. He warns that environmental contaminants will continue to present a risk to the world's marine mammals, even well into the twenty-first century.

In Part 3, Ozturk and Ozturk introduce the characteristics of the Black Sea in terms of its geology, oceanography, and biodiversity. They discuss threats to the biodiversity resources in the Black Sea posed by the fisheries industry, eutrophication, pollution by hazardous chemicals and alien species, and coastal degradation. Bastida *et al.* report characteristics of marine biodiversity of the south-western Atlantic Ocean with the due consideration of oceanographic features, and discuss the main environmental problems due to overfishing, increased contamination, and human use of inshore habitats.

Summary of key findings

Anthropogenic impacts

The various contributors to this volume repeatedly emphasize that we need to understand better the cycling of materials and nutrients in the oceanic systems. For example, Doumenge (Chapter 2) makes the point that fisheries the world over are adversely impacted by over-exploitation of stocks – resulting in bankruptcy for fisheries industries in many places. Ironically, more aggressive fishing approaches to cope with the depleted stocks lead to worsening of the situation. A solution to these problems can perhaps be found through focused research, including modelling of marine systems as demonstrated in the Japanese examples in Chapter 4 by Handa.

Land-based sources of pollution, such as industrial effluents, untreated municipal sewage, and runoff from agricultural areas, are the biggest threats to the coastal areas. In China, for example, pollution monitoring has clearly shown that environmental pollution in the coastal zones in the vicinity of river mouths and sewage outlets is particularly severe (Zhou, Chapter 6). In the Bohai Sea this has led to a complete collapse of the local fishing industry and numerous cases of red tide. Monitoring of shellfish for pollutants in the Asian region confirms that the trends of pollutant input into the coastal area are continuing (Prudente *et al.*, Chapter 7).

Eutrophication in coastal waters and the occurrence of harmful algal blooms (e.g. red tides) are another poignant reminder of the human impact on coastal areas. Okaichi and Yamada (Chapter 3) show that eutrophication in the Seto Inland Sea is a result of excessive influx of nitrogenous and phosphorus compounds, together with other organic pollutants discharged from manufacturing plants and cities along the coast.

Enclosed and semi-enclosed seas provide a good opportunity to observe directly the impacts of land-based pollutants. The examples cited in this volume, such as the Seto Inland Sea, the Black Sea, and the Bohai Sea, clearly demonstrate the adverse impacts on species through pollution. These seas provide clear examples of the impacts of unwise fishing practices, where the local fish stock can be depleted beyond recovery. For example, the number of fish species available in the Black Sea for sustainable commercial use has gone down from 27 in the 1970s to six (Ozturk and Ozturk, Chapter 11).

Species at risk

A number of species are identified as being at risk from man-made pollutants and intrusion into their habitats. In this list, mammals are quite prominent as they typically sit at the top of the food chain. The species discussed in this volume include:

- pinnipeds: Baikal seals (*Phoca sibirica*); northern elephant seals (*Mirounga angustirostris*); northern fur seals (*Callorhinus ursinus*); California sea-lions (*Zalophus californianus*); harbour seals (*Phoca vitulina*); and Mediterranean monk seals (*Monachus monachus*)
- whales and dolphins: minke whales (*Balaenoptera acutorostrata* and *Balaenoptera bonaerensis*); striped dolphins (*Stenella coeruleoalba*); Dall's porpoises (*Phocoenoides dalli*); and La Plata dolphins (*Pontoporia blainvillei*)
- dugong (*Dugong dugon*) and sea otter (*Enhydra lutris*).

The impacts on these mammalian species highlight the status of marine pollution as well as serving as an indicator of potential adverse impacts on marine ecosystems and the human food supply.

Strategies for the success of the "Mankind and the Oceans" concept

Conservation of marine and coastal species and ecosystems is at the heart of the "Mankind and the Oceans" concept. Sustainable utilization of these "ecosystem services" for human society is central and essential for the successful implementation of this concept. A number of human activities, particularly where industry and commerce rely directly on marine resources, have to be redesigned in a manner such that the needs of the local communities are met and their traditions are preserved. This clearly needs the involvement of local communities in planning, development, and implementation of any management strategies.

Development of sustainable fisheries is perhaps the most important element of a successful "Mankind and the Oceans" paradigm. A number of approaches for controlling the fisheries sector have been implemented; these include regulations against overfishing through setting quotas and licensing, restrictions on gear, limitations on fishing seasons and areas, and so forth (Doumenge, Chapter 2). However, these approaches have met with only limited success, particularly in developing countries. One has to dig deeper to see the underlying problems of poverty, lack of alternative livelihoods, and, often, a lack of political will to enforce these measures. This means that a change in our approach towards marine resource management is needed. This could include newer concepts like marine ranching and farming, which are introduced by Doumenge in this volume.

Another key element deserving our attention is the interface between land and ocean: the coastal zones. Special attention must be paid to the management of coastal resources and the integrated coastal zone management approaches that have evolved over the years. Their successful implementation still requires the following elements (adopted from Doumenge, Chapter 2):
- involvement of local communities and presence of political will
- a well-defined planning horizon and time period
- close interlinkages with national development planning
- clear and easy-to-follow guidelines for resource managers
- clear institutional arrangements, including laws and legislation
- monitoring and assessment for success of strategies.

It is important to point out that the collective experience of coastal management, as described by Okaichi and Yamada (Chapter 3), can serve as a useful tool in the implementation of integrated coastal zone management approaches. Management and minimization of land-based sources of pollutant and nutrient influx to coastal areas must be a central element in such approaches.

Scientific research, coupled with comprehensive monitoring and assessment, is fundamental and essential to the success of any strategies devised for the management of marine and coastal resources. A number of key research issues have been identified by contributors to this volume. Most importantly, the need to understand the impact of pollutants and coastal development on various marine and coastal ecosystems is highlighted by many contributors. Incidences of mass mortality of marine organisms and occurrences of harmful algal blooms are of particular concern throughout the world. Research presented here suggests that significant changes to our lifestyle, and to industrial and agricultural practices, may be needed to counter these adverse effects in the future.

The international community can play a key role in developing relevant institutions, focusing the research agenda, and providing human and financial resources for implementation of proven strategies, as argued by Adeel and Uitto (Chapter 5). As the problems facing marine and coastal ecosystems are without any arbitrary boundaries, so must our actions be. Only by working together in partnership can we hope for success of the concepts that underlie the notion of "Mankind and the Oceans".

REFERENCES

Borgese, E.M. 1998. *The Oceanic Circle – Governing the Seas as a Global Resource.* United Nations University Press, Tokyo, Japan.

Korzoun, V.I. (ed.). 1978. *World Water Balance and Water Resources of the Earth*, p. 663. UNESCO, Paris.

Shiklomanov, I.A. 2000. Appraisal and assessment of world water resources. *Water International*, 25(1): 11–32.

Voituriez, B. and G. Jacques. 2000. *El Niño: Fact and Fiction*, IOC Ocean Forum Series, UNESCO, Paris.

Part I

Human activities related to marine life and management

2

Mankind belongs to the sea

François Doumenge

Introduction

Throughout history, societies have regarded the seas as an inexhaustible resource. The seas provide food from edible seaweeds, molluscs, crustaceans, turtles, fish, sea-birds, and marine mammals; salt to flavour meals and conserve products such as driftwood, hard shells, or bones for tools and handicrafts; precious substances like ambergris, red coral, mother of pearl, tortoiseshell, and sea ivory; and skins or furs from seals, sea-lions, and sea otters.

The seas also have provided routes for long-range transportation, at a cheaper price than by land. Sailing and rowing were the foundation of many halieutic societies from antiquity (Greeks and Phoenicians) through the Middle Ages (Vikings, Hanseatic League, Venetians and Genoans) to the present time (Inuit, Malayan sea-faring *badjaos*, Maldivians, Polynesians, and Micronesians).

Harbour facilities support the economic development of commercial and industrial port areas in many bays, estuaries, and deltas. At present 65 per cent of cities with populations above 2.5 million are located on the seashore. Soon, the world's coastal areas will be home to 70 per cent of the total human population. In such conditions, mankind depends on the conservation of the seas. Unfortunately, in past decades economic development has been a cause of major degradation of the coastal environment, including eroded dunes and deltas, natural areas destroyed by landfills, harbour dredging and seabed mineral extraction accelerating beach erosion and stressing the benthic communities with siltation, and the pressures of tourism, which result in the degradation of shores and adjacent coastal waters. Densely populated coastal areas and their intensive economic

9

activities generate serious pollution (oil spills and transportation hazards, littering, faecal contamination by unprocessed sewage waters, bioaccumulation of toxic metals and organic substances from industrial discharge, and so on).

Common use of shore and sea resources was a basic Roman law that led eventually to the concept of the modern state's right of appropriation. But competition for ocean resources supported by too-speedy modernization quickly led to a worldwide decrease in ocean productivity. Only through self-regulation can we return to a convenient level of renewable resources. The transfer of ocean conservation management from national regulation to international bodies is a necessary step towards saving our oceans, because the governance of the seas has already established some common rules for fisheries, navigation, and the protection of the coastal and marine environment.

Overfishing: The collapse of living resources

Throughout the world, coastal and oceanic fisheries are hurt by rapid and deep depletions of biological stocks, bringing economic bankruptcy to fishermen. Fishing harder brings even more disaster, both for stock conservation and for economic returns, and leads to widespread overfishing.

Biological overfishing occurs when fishing reduces the biomass to a level where recruitment and/or growth are adversely affected. Economic overfishing exists when a fishery generates no revenue because it employs an excessive level of effort. Economic overfishing does not imply biological overfishing, as the latter is not necessarily the result of excessive capacity. Probably the foremost problem in the management of the world's sea fisheries is their susceptibility to "overcapitalization" (Marr, 1976). This tendency is usually attributed to the "common property" character of marine resources, to which everyone is presumed to have free access.

Since the yield of any fishery is finite, under conditions of unlimited entry of production units the total catch will be divided among more and more vessels until individual profits begin to decline. When increased catches by more productive, enlarged fleets result in decreasing yield per fishing unit, the fishermen compete to maintain their profitability, wishing to redeem their expensive modern boats and fishing tackle. The usual way to bring such a burden under control is for those fishing companies unable to raise enough money as a consequence of their declining yield to file for bankruptcy. The only solution to this problem is to restrict open access to fisheries, which has to be curbed by national and international regulations and agreements. This was the standard experience of growing industrial fisheries in the sixteenth to the twentieth centuries, affecting herring in the north-eastern Atlantic, cod in the North European and North American periarctic seas, and whaling and sealing worldwide. This resource depletion led to the imposition of an international regulation. It was the same scenario after 20 years (1953–1973) of competition in the bonanza of the Bering Sea, in North-West Africa, and in Pacific Latin America. Realizing to what extent their stocks had been destroyed through large-scale industrial trawling and purse seining, as well as through other techniques such as the use of giant drift nets and long-lines, coastal states used the political pressure of the emergent international law of the sea to establish a 200-mile Exclusive Economic Zone, and extended international regulation to high-seas fisheries. But this has not been sufficient.

Another unfortunate situation arises when fierce competition leads to a struggle to extract a supposedly inexhaustible resource. In such a case the overcapitalization for new boats and fishing technologies generates an exponential growth of the yield, glutting the market quickly. A good example of such a collapse is the history of the rating for Mediterranean precious coral at the end of the nineteenth century as the consequence of the discovery of three offshore banks in front of Ciacca (south-west of Sicily). More recently (September 2000), huge frozen skipjack landings by the industrial purse seiners fleets induced a financial crash; the world market price fell to US$350 a ton, bringing down profits for the whole world fleet as the minimum economic price for an industrial skipjack purse seiner is US$550–600 a ton. To save their investment the leading fishing countries (Japan, South Korea, Taiwan, the USA, Spain, and France) agreed to stop fishing for several months and enforced a strict quota for the catch rate of their boats. This was not an intergovernment agreement but a strictly free industry action to save a huge amount of venture capital. Such a responsible fishery action impressed the market. Within six months the frozen skipjack price returned to its former profitability level.

Modernization of small-scale fishing operations has also put new pressure on shallow seas and coastal areas worldwide. The primary danger here is not the human population pressure, but the fact that entrepreneurs may introduce overefficient catching systems before knowledge of expansion potential becomes available. Uncontrolled modernization may generate an overfishing situation in communities where the same stocks have been used without depletion for years. That is the case in many Indo-Pacific communities, as well as throughout coastal Africa and Latin America. Such innovations as the modernization of canoes and small craft for better access to fishing grounds, synthetic fibre fishing gear, and smaller winches and power-blocks for greater catching capacity have all improved small-scale fisheries' productivity dramatically. We are critically in need of a control to the expansion potential of local fishing waters, where traditional methods are geared to low harvest rates but where catching capacity is subject to sudden intensification (Cordell, 1977; Nietschmann, 1974).

Policy saving renewable living resources

Protection of endangered species

International agreements and national regulation can protect endangered species. Whaling, sealing, and turtle-catching are prohibited worldwide (with a few exceptions for some traditional native communities). Industrial tuna purse seining must give dolphins and porpoises associated with the tuna schools a means of escape (Joseph, 1994). Industrial pelagic drift-net fishing is on the verge of total prohibition after a moratorium.

Regulations against overfishing

Quotas

Setting a quota for a fishery is frequently used as a management technique. One reason for this is its simplicity; the management authority merely announces the quota, then, once this quota has been met, the fishery closes for the remainder of the season. The problem with any quota scheme lies in its enforcement. The scheme is likely to drive prices higher, so that individual fishermen have a greater incentive to exceed their quota on the sly.

Licensing

In establishing a licensing scheme, the first problem is "what is to be licensed". Occasionally it is the fishing gear itself. More frequently, it is the fisherman. Usually, however, it is the boat. While the aim of the licensing scheme is to bring effort under control, neither men, gear, nor boat makes a perfect proxy for the effort, and problems arise. One such problem is "substitution". While authorities can license an isolated part of the fishing effort (number of fishermen, amount of gear, number and gross tonnage of vessels), fishermen will likely find it profitable to "substitute" uncontrolled inputs for controlled ones. This phenomenon has come to be known as "capital stuffing". When, for instance, gross tonnage is controlled, fishermen fit more powerful engines or hauling gear for larger nets. The end result is that the catching ability of licensed vessels increases.

Gear restrictions

Another common method of regulation is gear restriction. Certain kinds of fishing equipment are either banned or circumscribed in their use. Such restrictions include limits on engine power, boat length, number of hooks per line, net and mesh size, dredge size, and beam width; they can even ban the use of motorized vessels altogether.

Closed seasons

A closed season is a period of time during which fishing cannot take place. As usual under open access, the individual fisherman will not find it profitable to invest in the fish stock because he has no way of ensuring that he reaps the benefit of his patience. In such situations, a closed season during a period when the fish are in poor physical condition will likely result in an increase in the value (suitably discounted) of the catch that exceeds the temporary sacrifice. Note, however, that the regulation attacks the symptom, rather than the problem itself (which, once again, is open access).

In cases where the stock becomes particularly vulnerable only at certain times, a closed season during these periods is an effective way of reducing the overall catch, and it is generally at such times that closures are implemented. The drawback with such closures is that the periods of vulnerability are usually the best times to fish, from an economic viewpoint, because the fish are easy, hence cheap, to catch. To the extent that a closed season brings a fishery under control, it will do so by making it more expensive to fish, thereby altering the open access equilibrium. The biological conditions of a fishery may well improve following the introduction of a closed season, but the economic condition will not. If there is no other way to control fishing efforts, a closed season may make sense for the survival of the fishery. Where alternative management techniques are available, however, it probably makes sense to use them.

Closed areas

A closed area is a section of the sea where fishing cannot take place. The area may be closed for part, or all, of the season, and is generally introduced to protect young or spawning fish. Since fishermen in an open access fishery will capture fish at too early an age, the protection of a nursery area may increase the weight of the sustainable catch. The best results are achieved when immature fish are separated from adults. The greater the mixing of fish of different ages, the more expensive a nursery area will be, since more mature fish must be sacrificed to protect the immature. It will be an empirical matter to decide whether a nursery area is justified or not. One use of nursery areas might be to protect the immature fish of one species being taken as a by-catch of a fishery directed at another.

Enhancement of marine living resources

Sea ranching

Ranching capitalizes on resources already in the natural environment by introducing a juvenile stock that feeds upon a renewable non-commercial species (Doumenge, 1995). The basic requirement is a direct relationship between the biological cycle of the ranched species and the time parameters of the natural environment. This type of development depends on the carrying capacity of the marine environment. The sea-ranching process must be repetitive, and the final result is in proportion to the level of artificial seeding. Ranching operations depend upon the control of juvenile production (breeding stock, massive spawning, larval stage growth, and safe metamorphosis). Hatchery-nursery management requires an efficient and cheap approach. The release of young organisms is arranged in convenient grounds and seasons. The best species for sea ranching are anadromous: salmonids and sturgeons. After a trophic life in the marine environment, they return to coastal areas to reproduce. An easy and efficient fishery can be established by catching these reproductive schools in trap nets posted at estuaries and river mouths, or by fishing at sea with small craft and gear when the mature fish are approaching.

Until now, only North Pacific salmon (*Oncorhyncus keta*) ranches have yielded high returns, for northern Japanese, Far East Russian, and Alaskan fishing communities (in the case of the Japanese, a yearly releasing of 2,060 million smolts during 1989, 1990, and 1991 produced a return of 2.9, 3.3, and 3.8 per cent of fish for a total weight of 220,000 tons in 1993, 250,000 tons in 1994, and 280,000 tons in 1995 respectively (Sato and Ito, 1999)). British Columbia and the American Pacific states are restricting or prohibiting sea ranching for legal and professional reasons (Pitts, 1991). Transferring some North Pacific salmon species to the southern hemisphere (e.g. Chile, New Zealand, Tasmania) may also prove successful. In the case of Atlantic salmon (*Salmo salar*), usually a first-class fish for intensive farming, such developments make less sense, however, since homing-instinct problems related to relocation have been recorded in this species. Sea ranching and natural population coexistence raise problems for conservation (interbreeding, competition, pathologic transmission). Notwithstanding such restrictions, other ways may be investigated, such as domesticating in convenient grounds sedentary species, like royal sea bream (*Pagrus major*) in southern Japan, or scallops (*Patinopecten japonicus*) in the Sea of Okhostk (Ito, 1991).

Marine farming

In the 1960s new aquaculture methods, supported by growth in the world market, started a "blue revolution" (Doumenge, 1986, 1987). Progress in the domestication of many valued species, development of adequate technologies, plus a declining natural harvest, all warranted extraordinary developments in new aquaculture systems (Barnabé, 1990).

Marine farming builds new coastal landscapes; on the land side, fish and shrimp ponds have spread over wetlands, deltas, lagoons, and tidal flats, not only to replace salt marshes but also mangroves, dunes, and marsh embankment. Along the seaside, floating structures (cages, rafts, long-lines) fill many shallow bays, rias, and fjords, and make rapid progress offshore.

The best resources for marine aquaculture are at the lower level of the trophic chain, which produce several marketable products, including edible seaweeds (e.g. *Porphyra, Laminaria, Undaria pinnatifida, Gracilaria, Euchema*, etc.; Perez *et al.*, 1992), capitalizing on seawater nutrients such as bivalves (e.g. oysters, mussels, pearl oysters (*Pinctada*), scallops), sea cucumbers, sea snails and sea urchins, and fish that graze on the bacterian film or the algal turf (milkfish, mullets).

Unfortunately, this efficient way of harvesting extra productivity from the base of the marine food chain is often restricted either to local tastes (edible seaweeds for the Eastern Asians, milkfish for the Indo-Malayan peoples) or by a too-low market price (mussels, mullets). On the other hand the more valuable marine species, penaeid shrimp and carnivorous fish (salmon, sea-bass, sea-bream, yellowtail, flatfish such as turbot), need not only clean seawater with specific temperature levels (solar energy giving a free input for the growth metabolism) but also enough protected space and volume to support very high densities. Aquaculture can be a way to value and protect endangered species before over-fishing occurs, as was the case with the bluefin tuna (Doumenge, 1996).

Aquaculture requires careful seawater management and free circulation for clean water. Such requirements generate use conflicts over space. A good example of such a situation is the People's Republic of China, where huge coastal densities of humans compete for the use of coastal space (Yu, 1994).

Developing new habitats

Building new habitats that can be used to protect and enhance existing resources also compensates for resource depletion resulting from pollution, environmental degradation, and overfishing (Seaman and Sprague, 1991). "Artificial reefs" are bio-ecological vessels that encourage new production by recycling biochemical energy within eutrophic areas, and generate halieutic enhancement via multiple effects: reduction of juvenile mortality; protection of breeding grounds and nurseries; and concentration of reproductive stocks.

Different strategies support the building of large artificial reefs. In the USA, for example, the rehabilitation and development of fishing grounds are oriented to the maintenance of sport fisheries (Edwards, 1990) and the reuse of out-of-commission oilrig platforms (Reggio and Kasprzak, 1991). In Italy, multi-purpose concrete structures act as bio-ecological mechanisms able to enhance the fishery biomass and support a sessile community dominated by marketable filter feeders (Bombace, 1995).

In Japan, such vessels conserve resources for the support of traditional fishing villages. Instead of merely increasing catches in natural fishing grounds, however, a 19-year programme (1976–1994) was launched to create entirely new fishing grounds or habitats where there had been none (Grove *et al.*, 1994). These habitats were stocked with juvenile, post-larvae crustaceans and fingerling fishes, mass produced by an efficient network of hatchery-nurseries to the public service (national or departmental), or to the fishermen's cooperatives. Approximately US$8 billion was spent by the end of the programme (March 1994). Looking some years ahead such a policy return is impressive. With a stable catch rate the Japanese small-scale fisheries maintain higher living standards for the fishermen than in any other countries.

For the year 2000, the total landings value for less than 20-ton boats reached US$4 billion, representing 40 per cent of the total value for national landing with an equal 10 per cent for each small-scale fishing boat (less than 3 ton, 3–5 ton, 5–10 ton, 10–20 ton).

Coastal and shallow seas: Common use or tenure

In many countries the greatest threat to sustainable marine resources is instability and poverty, which force individuals and communities into short-term decisions to exploit resources at an unsustainable level. Such pressure, coupled with exploding populations and technological advances, is particularly detrimental when the marine environment is a "commons", where resources are open to anyone willing to exploit them. Damage can be controlled, however, if coastal communities manage the contiguous sea as a tenure as happens on the shore and adjacent lands.

Many traditional groups have a long history of "private use management", developed as a specific way of life for their society: Pacific islanders on mangrove swamps, coral reefs, and lagoons; Indo-Malayans on lagoons and shallow bays; American Indians, Paleo-Siberians, and Inuit along fjords, straits, and island channels (Ruddle, Hviding, and Johannes, 1992). In this context the use of the term "tragedy of the commons" in the sense of "a tragedy of open access" with the addition of new competitors for space, services, and natural resources is not inappropriate.

Open access generates conflicts, disorder, and dysfunction between individuals and groups, spoiling the future with the misuse of new technologies and causing irreversible damage. In contrast, by covering tidal flats and shallow seas with rights granting professional groups exclusive use and responsibility for a certain area, for their exclusive profit, two competing goals can be reconciled: the conservation of natural renewable resources and the integration of traditional and new technologies for common benefit.

An example of this is the Japanese coastal management system for small-scale fisheries and aquaculture, which grants exclusive rights for specific sections of public sea grounds and waters (Asada, 1973). Consequently, they are regarded as "rights *in rem*" (property rights), and the same provisions concerning land are applied to them in legal transactions, except for certain restrictions on lending, transfer, and mortgage. The foundation of such an approach is the national "Old Fisheries Law" of 1901, which recognized and legalized traditional institutions and practices of the coastal communities' ways of life. At the turn of the twentieth century, a time of rapid social and economic transformation, this legislation prevented fishing conflicts and protected marine resources, while at the same time giving the socio-professional group of fisherman's cooperatives (*Kumiai*) and private entrepreneurs responsibility for maintaining or changing their traditions. This law was appropriate to the new circumstances of reconstruction after the Second World War, in 1949, and in 1962, at the beginning of the new economic boom.

The current "fishing rights" system includes the following three types of management.

- Common fishing rights, granted only to *Kumiai*, which take responsibility for productive operations, marketing, and the economic and social support of the group. At "general meetings", *Kumiai* take all decisions on the use of the grounds covered by their rights, which can be rented or even sold to other users for development or services.

- Set-net fishing rights, which allot space for traps more than 27 m deep and place the net wall (tail) on the shore. First preference for granting set-net fishing rights is given to *Kumiai*, and second to fishing entrepreneurs living in coastal areas.
- Demarcated fishing rights, giving permission to engage in private aquaculture. Priority is given to *Kumiai*, whose members must operate according to the specific regulations decided by the group. The right to enter as a private tenant can also be granted to individual applicants who fulfil certain requirements such as qualification, residence, and contributions to the group. With such a simple, strict, but adaptable system, Japanese coastal development can be reconciled with the sustainable management of resources, as well as with balancing basic tradition with the pressures of the market and technological modernization (Simard, 1989).

Basic concepts for a coastal management action plan

Carrying capacity

For the purposes of managing an area, *carrying capacity* can be defined as the threshold level of use above which there is overloading. This is a simplistic definition, but it is often difficult to establish precisely when overloading occurs. Carrying capacity measures the amount of use beyond which impacts exceed acceptable levels specified by evaluative standards (Brotherson, 1973; Cordell, 1977). Carrying capacity identifies a number for one management parameter: use level. It assumes a fixed and known relationship between use level and impact parameters, and the capacity will change if other management alters that relationship. Capacity will also change if management objectives are altered or if user populations change radically.

Shelby and Heberlein (1984) distinguish between the following four types, differentiated by decisions about which kinds of impact are important.

- *Ecological capacity* (ecosystem parameters). Ecological capacity can be used to indicate the general productivity of ecosystem performance. In such a concept the carrying capacity is the asymptote population biomass supported by an ecosystem, under the limitation of food, shelter, and so on, and under the effects of predation and exploitation (Kashiwai, 1995). The most critical point will be whether or not the ecosystem can have any semi-equilibrium state and asymptote biomass under the regional shift in physical environment.
- *Physical capacity* (space parameters). Physical capacity is the amount of space potential that can effectively be managed according to local, national, or

international boundaries. At present it must take into account the shift affecting the use of the seas concept from open access to restricted areas for political, economic, or conservationist reasons. The physical capacity belongs to the nature and dynamics of the fishing ground. Coastal morphology induces specific biota: sandy beaches, muddy flats, mangroves, and coral reefs. Some characteristic areas like lagoons and lidos broken with passages, rias and fjords, estuaries and deltas, and atolls support large life concentration by seasonal trophic or genetic migrations. In the same view oceanic currents generate high pelagic productivity. The physical capacity of an area could be changing over short periods, such as during El Niño/La Niña (Doumenge, 1999) or during North Atlantic and North Pacific oscillations, rather than during long-term global change such as polar seas warming. In such circumstances beyond one's control it is necessary to change the management and the life-support system of the fishing community.

- *Facility capacity* (development parameters). Facility capacity involves manmade improvements intended to handle the needs of fisheries including the infrastructure, such as ports and harbours with all their storage, transport, and processing facilities. With the global spread of new technologies the present dilemma is not motorized versus non-motorized craft, or natural versus synthetic fibres, but access to GPS, real-time information, and operational research capacities. Formative and administrative staffs are also included in this category because they facilitate use.

- *Social capacity* (experience parameters). Social capacity refers to impacts that impair or alter human life. Social carrying capacity is the level of use beyond which experience parameters exceed acceptable levels specified by evaluative standards. Impact parameters focus on the number, type, and location of encounters with other human groups, and on the way these encounters affect marine resource use. Social capacity has traditionally been difficult to determine, primarily due to the difficulty in establishing evaluative standards. To establish social carrying capacity, there must be:

- a known relationship between use level or other management parameters and experience parameters
- agreement among relevant groups about the types of fishery or sea farming experience to be provided
- agreement among the relevant groups about the appropriate levels of the experience parameters.

To enhance social capacity, the first objective should be to promote education, explaining to fishermen and decision-makers the common purpose underlying policies and management.

Limits of acceptable change

Impacts should be identified, whether they are social, environmental, or physical. This may be done by conducting field surveys to assess the levels of vegetal and animal stocks; to identify the ecological systems' relationships; to quantify the damage caused by overcrowding and overexploitation; to look at levels and frequency of congestion affecting the natural and socio-economic systems (i.e. red tides and eutrophication, sand beach and dune destruction, dysfunction of public services like water systems, sewage and garbage dumping, public transportation blockages, etc.); and to record conflicts between various user groups (Johnson and Pollnac, 1989; De Voe and Pomeroy, 1992).

From these surveys, problem areas can be identified. This involves quantifying the capacity for acceptable change, looking at a gap or critical level where it will be impossible to control the alteration of the environment or the perturbation of the economic and social way of life in the area (Figure 2.1) (Marion, Cole, and

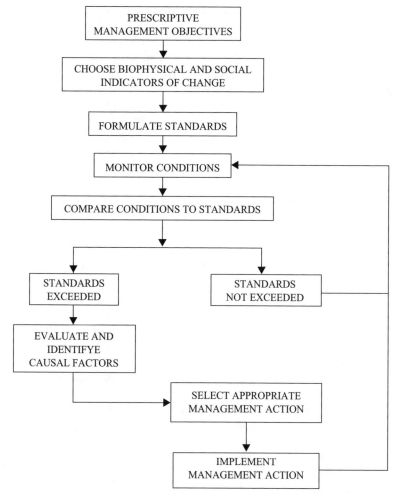

Figure 2.1 The simplified version of the limits of acceptable change (LAC) planning process
Source: Marion, Cole, and Reynolds, 1985

Reynolds, 1985). Management decisions must be based on scientific criteria for the natural factors, and on the area's social and economic norms for the human factors. Once the impacts have been identified and standards set and converted to acceptable user levels, the overall carrying capacity must be determined at the lowest limit of all these factors. The overall carrying capacity is not always based on the limitations of the natural resource but, at times, on the limitations of the social or economic aspects, or on the capacity of the physical equipment available.

Integrated coastal zone management

Responsibility for coastal conservation should be part of a national planning policy that could regulate the spatial repartition in accordance with the carrying capacity of the natural and socio-economic environment and within the limits of acceptable change. A general agreement is emerging using the approach of "integrated coastal zone management" (ICZM) (Thia-Eng, 1993). The goal of integrated coastal management is to produce the optimal mix of products and services from the coastal ecosystem over time, with "optimal" being the mix that results in maximum social benefit. The political process usually defines the mix. Since the interests and priorities of society change over time, so does the mix (Figure 2.2).

Some fundamental concepts provide the basis for common ICZM criteria for all places and circumstances. Kenchington and Crawford (1993) suggest that the following elements are required.

- A dynamic goal or vision of the desired condition of the oceanic or coastal area, and the integration of human use and impact for a period significantly longer than conventional economic planning horizons, say, 25 or 50 years.
- Broad national objectives: commonly agreed upon aims of common purpose to which policies and management are directed. It can be expected that some of the objectives will be mutually contradictory; provided, however, that their achievement is subject to the overarching purpose, each is constrained with respect to the others, and there is a basis for mediating or arbitrating disputes. For regional and local plans, progressively more detailed objectives that remain consistent with the national objectives are usually required.
- Guiding principles for managers or statutory decision-makers with discretionary powers for planning, granting approvals, or making changes to the purpose or extent of use and access.
- A strategy, commitment, and resources for the objectives to be met through detailed day-to-day management that may involve several agencies and the community.
- Clear, legally based identification of authority, precedence, and accountability for achievement of the strategy in relation to any other legislation applying to the area in question.
- Performance indicators and monitoring to enable objective assessment of the extent to which goals and objectives have been met.

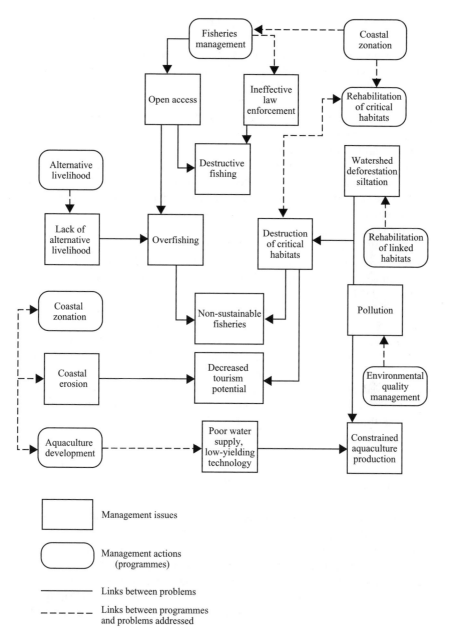

Figure 2.2 Linkages between management action and management issues, Lingayen Gulf
Source: Thia-Eng, 1993

- Above all, there must be political, administrative, and stakeholder will and commitment to implement the strategy.

Following these recommendations, it was agreed in an expert meeting (Charleston, USA, July 1989) that an ICZM programme should have the following five attributes (Sorensen, 1993).

- It is a process that continues over a considerable period of time.
- There is a governance arrangement to establish policies for making allocation decisions, and if the programme is implemented, a governance arrangement for making allocation decisions.
- The governance arrangement uses one or more management strategies to rationalize and systematize the allocation decisions.
- The management strategies selected are based on a system's perspective, with recognition of the interconnections among coastal systems.
- It has a geographic boundary that defines a space which extends from the ocean environment across the transitional shore environments to some inland limit.

Problems for putting such policy in action flow from three types of deficiency and failure, or a complex combination of these coastal zone management strategies (OCDE/OECD, 1993):

- policy deficiencies, where policies are absent or contradictory, leading to sub-optimal results and reduced net social benefits
- intervention deficiencies, where governments have decided to intervene through government-led coastal zone management, but where this intervention fails to achieve all or any of the objectives set, for a variety of reasons
- market failure, where prices and resource allocation decisions fail to reflect the real value of coastal zone resources, or where markets are distorted by taxes and subsidies such that net social benefits are lower than would otherwise be the case.

Conclusion

From this short review of the main resource management tools, links with broader issues should be apparent. Conservation, rehabilitation, and sustainable use of shore and coastal benthic resources cannot be achieved without consideration to pelagic and large marine ecosystems. In addition, protection of sand dunes, lagoons, estuaries, deltas, mangrove swamps, and coral reefs cannot be effective without a strong control of adjacent continental lands, where forest clearing, agricultural intensification, and industrial and urban growth result in coastal pollution and natural equilibrium perturbations.

Ideally, integration should extend to coordinate the management of marine and terrestrial areas in the coastal zone and beyond, covering both continents and oceans. This necessity springs from the variety of legal and jurisdictional systems, and of numerous management authorities, that have been established separately and with quite different mandates. How then to avoid conflicts, which usually result in degradation and overuse? It is a strong priority to define clearly responsibilities for management, and for institutional and administrative mechanisms to ensure coordination via a general scheme built on objective and scientific criteria.

As was made clear in the 1992 Rio Conference, advancing technologies are not in opposition to the conservation of the environment and to the maintenance of balanced socio-economic progress; but the support of clear concepts of responsibility is required. A dynamic goal of the integration of the human future with the oceanic world must be planned for a period of 25–50 years, significantly longer than conventional economic policies. Any action concerning the seas and their coastal border is a process that continues over considerable time.

REFERENCES

Asada, Y. 1973. License limitation regulations: The Japanese system. *Journal of the Fisheries Research Board of Canada*, 30(12): 2085–2095.

Barnabé, G. 1990. *Aquaculture*. Ellis Horwood (ed.), London, 2: 1104pp.

Bombace, G. 1995. Le barriere artificiali nella gestione razionale della fascia costiera italiana. *Biologia Marina Mediterranea*, 2(1): 1–14.

Brotherson, D.I. 1973. The concept of carrying capacity of countryside recreation area. *Recreation News Supplement*, 19: 6–11.

Cordell, J. 1977. Carrying capacity analysis of fixed-territorial fishing. *Ethnology*, 17: 1–24.

DeVoe, M.R. and R. Pomeroy. 1992. Use conflicts in aquaculture: A worldwide perspective on issues and solutions. *World Aquaculture*, 23(2): 13–35.

Doumenge, F. 1986. La révolution aquacole. *Annales de géographie*, No. 530, pp. 445–482.

Doumenge, F. 1987. La révolution aquacole. *Annales de géographie*, No. 531, pp. 29–86.

Doumenge, F. 1995. L'interface pêche/aquaculture, coopération, coexistence ou conflit. *Norois*, 42(165): 205–223.

Doumenge, F. 1996. L'aquaculture des thons rouges. *Biologia Marina Mediterranea* (Atti del XXVI Congresso della Societá italiana di biologia marina, Sciacca, Italie, 20–27 May 1995), 3(1): 258–288.

Doumenge, F. 1999. L'oscillation australe ElNino (ENSO): anomalies de l'hydroclimat et consequences. In: *Biologia Marina Mediterranea* (Atti del XXIX Congresso della Societá italiana di biologia marina, Ustica, Italy, 15–20 June 1998), 6(1): 1–51.

Edwards, S.F. 1990. An economic guide to allocation of fish stocks between commercial and recreational fisheries. *NOAA Technical Report*, NMFS/94.

Grove, R.S., M. Nakamura, H. Kakimoto, and C.J. Sonu. 1994. Aquatic habitat technology innovation in Japan. *Bulletin of Marine Science*, 55(2–3): 276–294.

Ito, H. 1991. Successful HOTAC methods for developing scallop sowing culture in the Nemuro district of East Hokkaido, northern Japan. In: *Proceedings of the Seventeenth U.S.–Japan Meeting on Aquaculture, Ise, Mie Prefecture, Japan*, 16–18, October 1988, R.S. Svrjcek (ed.), NOAA Technical Report/NMFS 102 Marine Ranching, pp. 107–116.

Johnson, J.C., and R.B. Pollnac. 1989. Introduction to managing marine conflicts. *Ocean and Shoreline Management*, No. 12, pp. 191–198.

Joseph, J. 1994. The tuna-dolphin controversy in the eastern Pacific Ocean: Biological, economic, and political impacts. *Ocean Development and International Law*, 25: 1–30.

Kashiwai, M. 1995. History of carrying capacity concept as an index of ecosystem productivity (review). *Bulletin of the Hokkaido National Fisheries Research Institute*, No. 59, pp. 81–101.

Kenchington, R. and D. Crawford. 1993. On the meaning of integration in coastal zone management. *Ocean & Coastal Management*, 21(1–3): 109–127.

Marion, J., D. Cole, and D. Reynolds. 1985. Limits of acceptable change: A framework for assessing carrying capacity. *Park Science*, 6(1): 9–11.

Marr, J.C. 1976. Fishery and resource management in S.E. Asia. *Resources for the Future Program of International Studies in Fishery Arrangements*. No. 7. Washington, DC.

Nietschmann, B.O. 1974. When the turtle collapses, the world ends. *Natural History*, 83: 34–43.

OCDE/OECD. 1993. Gestion des zones côtiéres: Politiques intégrées. *Editions de l'OCDE*, Paris, p. 144.

Perez, R., R. Kaas, F. Campello, S. Arbault, and O. Barbaroux. 1992. La culture des algues marines dans le monde. *IFREMER*, No. 5, p. 614.

Pitts, J.L. 1991. The use of aquaculture enhancement of the common property fishery in Oregon, Washington, and Alaska. In: *Proceedings of the Seventeenth U.S.–Japan Meeting on Aquaculture, Ise, Mie Prefecture, Japan*, 16–18 October 1988. R.S. Svrjcek (ed.), NOAA Technical Report/NMFS 102 Marine Ranching, pp. 117–121.

Reggio, V.C., Jr. and R. Kasprzak. 1991. Rigs to reefs: Fuel for fisheries enhancement through cooperation. *American Fisheries Society Symposium*, No. 11, pp. 9–17.

Ruddle, K., E. Hviding, and R.E. Johannes. 1992. Marine resources management in the context of customary tenure. *Marine Resources Economics*, 7(4): 249–273.

Sato, K. and J. Ito. 1999. Gendai Sake jijo Niju oku o horyu no jidai. *Fish Culture*, 2: 98–105.

Seaman, W., Jr. and L.M. Sprague. 1991. *Artificial Habitats for Marine and Freshwater Fisheries*. Academic Press, Inc., San Diego, 285pp.

Shelby, B., and T.A. Heberlein. 1984. A conceptual framework for carrying capacity determination. *Leisure Sciences*, 6(4): 433–451.

Simard, F. 1989. Japon: La pêche côtiére, mutations technologiques et conséquences socio-économiques." *Equinoxe*, No. 27, pp. 25–33.

Sorensen, J. 1993. The international proliferation of integrated coastal zone management efforts. *Ocean and Coastal Management*, 2(1–3): 45–80.

Thia-Eng, C. 1993. Essential elements of integrated coastal zone management. *Ocean and Coastal Management*, 21(1–3): 81–108.

Yu, H. 1994. China's coastal ocean uses: Conflicts and impacts. *Ocean and Coastal Management*, 25: 161–178.

3

Environmental management of enclosed coastal seas

Tomotoshi Okaichi and Machiko Yamada

Introduction

Prior to the UN Conference on Environment and Development – the so-called "Earth Summit" – held in Rio de Janeiro in June 1992, the General Assembly of the UN Conference on Environment and Development (UNCED) in 1989 pointed out in Article 12 (2) of Annex I (1989) that "protection of the oceans and all kinds of seas, including enclosed and semi-enclosed seas, and coastal areas, and the protection of rational use and development of their living resources" is a major concern in maintaining the quality of the earth's environment, and especially in achieving environmentally sound and sustainable development in all countries.

EMECS activities from 1990 to the future

In recognition of the importance of enclosed seas, international conferences on EMECS have been held since 1990, with Kobe hosting the first gathering of national and local government representatives, private companies, and researchers. Subsequently four conferences have been held in four countries and in each conference the following declarations and recommendations were adopted.

From 3 August to 6 August 1990, the First International Conference on the Environmental Management of Enclosed Coastal Seas (EMECS '90) was held in Kobe, Japan, sponsored mainly by the Governors' and Mayors' Conference on the Environmental Conservation of the Seto Inland Sea. The conference was attended by 1,238 people from 42 countries. Scientific sessions were aimed at encouraging public interest in the environmental conservation of enclosed seas. A total of 143 presentations were made on the following five topics.

- The present state of environmental pollution in enclosed coastal seas, and measures for environmental protection.
- Ecological systems and fisheries resources in enclosed coastal seas.
- The appropriate use of enclosed coastal seas.
- Management and administration of enclosed seas.
- Enclosed coastal seas and human activities.

The unanimous adoption of the Seto Inland Sea Declaration, which stressed environmental management and the appropriate use of the world's enclosed seas as urgent issues internationally, was concluded in the conference.

Based on the Kobe Declaration the following activities constituting EMECS were carried out.

- The Research Institute for the Seto Inland Sea was set up in March 1992. Currently about 470 members are registered and working in universities, national and prefectural institutes, and the private sector. NPOs working for the environmental conservation around the Seto Inland Sea are also admitted into the Institute.
- The Hyogo Prefectural Government has been engaged in developing an international network among researchers. The network organizes each conference in tandem with organizing committees.
- At the Second EMECS Conference in Baltimore, held from 10 November to 13 November 1993, more than 500 attendants from 40 countries participated. The establishment of the International EMECS Center in Kobe, Japan, was proposed and unanimously accepted by participants. The declaration of the conference includes many important suggestions for further activities in EMECS, which are shown in Figure 3.1 (Declaration in Baltimore).
- The Third EMECS Conference was held in Stockholm from 10 August to 15 August 1997, in cooperation with the Stockholm Water Symposium. The theme of the conference was "With River to the Sea – Interaction of Land Activities, Fresh Water and Enclosed Coastal Seas". The conference was a great success, stimulating further study and facilitating interaction between freshwater and marine scientists to promote a holistic perspective. The recommendation adopted at the conference stressed the following principles: pursuing a holistic approach; improving understanding; developing an active dialogue; and acting locally – thinking regionally.
- The Fourth EMECS Conference was held in cooperation with the fourth Medcoast (International Conference on the Mediterranean Coastal Environment) in Antalya, Turkey, from 9 November to 13 November 1999, with the theme "Land-Ocean Interaction: Managing Coastal Ecosystems". In this conference, about 250 presentations were made. Throughout these international conferences, the importance of managing enclosed coastal seas was discussed, and the activities of organizations, both official and scientific, attracted the interest of participants.

At the Second EMECS Conference in Baltimore, major committees for Chesapeake Bay were established (Figure 3.2). The institutional framework for North Sea policies in the Netherlands was established at the Third EMECS Conference in Stockholm (Figure 3.3) (Vallejo, 1994).

We affirm that EMECS is dedicated to the following principles:

- facilitating the international exchange of scientific information, including advances in research and modeling on coastal phenomena

- fostering understanding among policymakers and researchers of the motivations and interests of citizens which are essential to the implementation of sound policy

- improving communication and cooperation across the increasingly important science-policy interface

- building upon common commitment to protect coastal seas because of their importance as places of physical beauty and cultural and historic meaning

- providing a venue for change of technology useful to solving problems of coastal seas

- pursuing new approaches to governance informed by our concern for ecosystems that cross the jurisdictional boundaries that mankind has imposed.

Figure 3.1 Declaration of Principles, EMECS '93 Baltimore

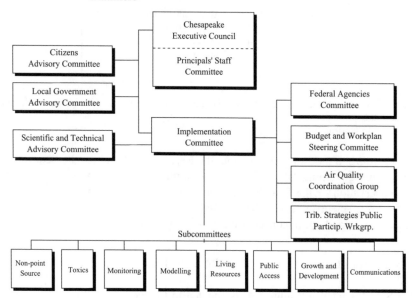

Figure 3.2 Major committees in the Chesapeake Bay programme

Figure 3.3 Institutional framework for the development of North Sea policies in the Netherlands
Source: van Hoorn, 1989

According to the Kobe Declaration, the Research Institute for the Seto Inland Sea plays a governing role, holding study forums every year on the environmental conservation of the Seto Inland Sea with contribution and participation by prefectures, scientists, and citizens. Measures to prevent outbreaks of red tide have been carried out in concert with fishermen and residents.

Environmental management in the Seto Inland Sea, Japan

The Seto Inland Sea (Figure 3.4) is situated in the western part of Japan and is famous for its beautiful landscape – including as many as 700 islands – and is designated a national park (Okaichi and Yanagi, 1997).

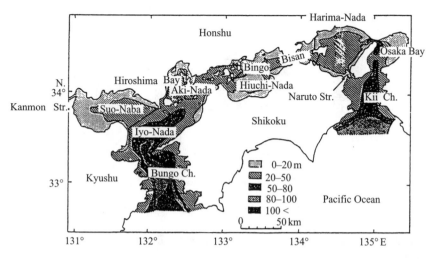

Figure 3.4 The Seto Inland Sea, Japan, and its topography

Its length is about 450 km, its width fluctuates between 15 and 55 km, and its average depth is about 37 m. The sea occupies about 22,000 km² and has three channels that open to the Pacific; it is typical of Japan's enclosed seas. The fisheries and aquaculture are prosperous and now they provide approximately 300,000 and 400,000 tonnes of products a year, respectively. On the other hand, the coastal area is one of the most industrialized in Japan, with a population of 3.5 million. Industrial production reaches 90,000 billion yen, so there is serious potential for contamination. Since 1955 there have been many serious water pollution problems around the Seto Inland Sea. Pushed by the heavy pollution occurring in many places in Japan, the central government issued the "Public Nuisance Countermeasures Basic Law" in 1967, which was later revised as the "Environment Basic Law" in 1994. Harmful red tides in the Seto Inland Sea have often occurred on a large scale after 1957 and brought losses of more than 20 billion yen over a 20-year period. The aquaculture, especially, was severely damaged.

In 1971 governors of prefectures and mayors of big cities located along the coast of the Seto Inland Sea hosted a conference for the conservation of the sea and presented the Seto Inland Sea Charter on Environmental Protection. The Governors' and Mayors' Association for the Environmental Conservation of the Seto Inland Sea was established in 1976 and now 13 governors and 13 mayors host annual conferences to discuss the general problems related to environmental management. The Environment Agency of the central government was established in 1971 and the interim Law for the Conservation of the Environment of the Seto Inland Sea was issued in 1973, followed by amendment to the permanent Law Concerning Special Measures for Conservation of the Environment of the Seto Inland Sea in 1978. In accordance with the law the national government

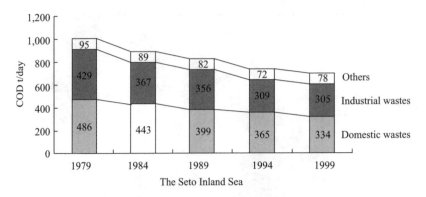

Figure 3.5 Trend of COD load in the Seto Inland Sea

issued the basic plan for the conservation of the environment of the Seto Inland Sea.

The main activities based on the measures recommended by the Environmental Agency (1993) are as follows.

- Establishment of environmental conservation plans in each prefecture.
- Permission for installation of specified facilities. According to the volumes and qualities of the polluted water, a licensing system was introduced to build new facilities. Five hundred facilities were given permits by the related prefecture in 1991.
- Enforcement of area-wide total water pollution control. With the enforcement of the control in 1974, COD (chemical oxygen demand) loading was expected to decrease by half in three years. The third series of area-wide pollutant load controls was achieved in 1994 (Figure 3.5). Total loading of COD decreased to about 760 tonnes in 1995 from 1,600 tonnes in 1970.
- Guidance for reduction of inflow of nitrogenous and phosphorus compounds. Prefectural guidance has been provided on the reduction of inflow of phosphorus compounds to the Seto Inland Sea by the Environment Agency since 1980 and guidance for nitrogen reduction was issued in July 1994.
- Conservation of natural sea coasts.
- Consideration for environmental conservation in reclamation. For permission for reclamation, governors of related prefectures request full consideration to be given to the special features of the Seto Inland Sea.

Countermeasures for eutrophication and red tide outbreaks

Generally speaking, red tides occur as a result of eutrophication. Eutrophication in the Seto Inland Sea was promoted by the inflow of nitrogenous and phosphorus compounds, together with other organic pollutants discharged from manufacturing plants and cities along the coast. The amount of pollutants in 1972 was reported to be 1,345 tonnes of COD per day, 470 tonnes of nitrogen, and 32 tonnes of phosphorus. In addition – unbelievable now – nearly 3,000 tonnes of untreated human excreta were discharged every day into the offshore water area prior to March 1973.

Red tides in the Seto Inland Sea were exacerbated by the appearance of species belonging to Raphidophyceae, *Chattonella*, in 1969. Since then mass mortalities in cultured yellowtail fish occurred from 1970 up to the present. Recently *Gymnodinium mikimotoi*, a dinoflagellate, which at first appeared in Tokuyama Bay, Yamaguchi Prefecture, in 1957, has again invaded the Seto Inland Sea. Since 1995 another invader, a dinoflagerate, *Heterocapsa circularisquama* attacked pear oyster and edible oyster culture fields, causing severe damage.

In 1972 a large-scale red tide due to *Chattonella antiqua* attacked the fish culture fields in Harima Nada, an eastern area of the Seto Inland Sea covering 3,426 km^2, and 14 million yellowtail fish and other fish were killed. The losses amounted to 7.1 billion yen in four prefectures. Following the accident many institutions of the central government and prefectures promoted comprehensive environmental surveys of the Seto Inland Sea.

Research on red tides after 1970 produced many results, which contributed to the prevention of the outbreaks of red tides and damages to fisheries. Studies on red tides in the Seto Inland Sea are reviewed by Okaichi (1997).

As shown in Figure 3.6, industrial production of the prefectures along the coast increased markedly after 1976, amounting to almost 90 trillion yen today. By contrast, the case numbers of the outbreak of red tides in the Seto Inland Sea, which were 326 in 1976, markedly decreased to about 100 cases per year since 1990.

The decrease in the outbreaks of red tides in the Seto Inland Sea can be attributed mainly to the following activities.

- Administrative guidance based on the Law Concerning Special Measures for Conservation of the Environment of the Seto Inland Sea, which controls eutrophication and regulates land reclamation.
- Installation and improvement of industrial and domestic wastewater treatment facilities.
- Progress in aquacultural techniques, including the use of pellets for fish diets and other feeding methods.
- Diffusion of education on the conservation of the environment of the Seto Inland Sea among the residents along the coast.
- Applications of the results of the scientific and technical researches.

The progress of the environmental management of the Seto Inland Sea is evident by the decrease in the outbreaks of red tides. However, environmental management has been insufficient, as land reclamation still continues and water quality has not yet reached the levels stipulated by the Ministry of Environment. More detailed planning for the environmental conservation of the Seto Inland Sea is still a requirement.

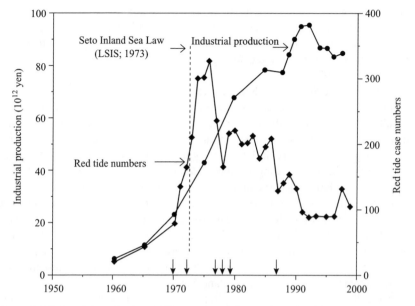

Figure 3.6 Industrial production (trillion yen) and the number of red tide occurrences
Note: Vertical arrows indicate years of fish mass mortality.

Successful improvements of the aquatic environment of Dokai Bay, Kitakyushu City

Dokai Bay, located in the northern area of Kyushu, is surrounded by heavy industries such as ironworks, docks, and chemical and food industries (Yamada *et al.*, 1990, 1998). The bay is narrow and long, with a width of less than 1 km and a total length of 13 km (Figure 3.7). Owing to untreated wastewater from the industrial complexes on the coast, the seawater and bottom sediment were polluted and fisheries' yields dropped by half between 1928 and 1932. The pollution became severe after the Second World War, and Dokai Bay became known as the "Dead Sea". Until 1965 the pollution in the bay was a problem only for the fishermen. But in 1965, the residents of the inner bay complained to Kitakyushu City about offensive odours due to the heavy pollution of the bay. Journalists also took up the issue of pollution and highlighted it as a social problem.

Kitakyushu City first carried out surveys of the water qualities of the bay in 1966 in response to the demands of the citizens. Dissolved oxygen at the three metre depth-mark in the inner and middle part of the bay was 0 mg/l. In 1969, COD values were as high as 74.6 mg/l and cyanide and arsenic were as high as 0.64 and 0.15 mg/l, respectively. These figures provided clear reasons for the Dokai Bay being called the "Dead Sea". In 1969 the total effluent discharge from

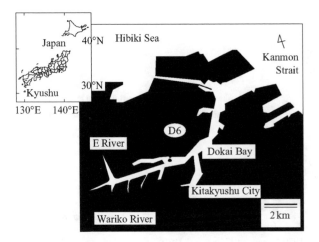

Figure 3.7 Location of sampling station D6 in Dokai Bay

land was thought to be 4,080,000 tonnes a day and COD 233 tonnes a day. Twenty-two big industrial plants were responsible for 98.5 per cent of total effluents and 97.3 of the COD loading.

After these investigations, sewage works were improved and dredging of the polluted seabed was carried out. Since then, regulations for industrial wastewater have been stipulated, first under the guidance of the Economic Planning Agency, then by the Environmental Agency of the city, including rules for the control of industrial waste, the construction of sewage systems, and dredging of polluted sediments. Along with these clean-up operations, negotiations between Kitakyushu City and industries have continued from 1970. Since 1968 each company has developed new water treatment systems and improved the treatment plants. The subsequent success of the Dokai Bay programme is attributed to cooperation between residents, scientists, industries, and administrative bodies in the city. The water quality of Dokai Bay has improved since 1971 (Figure 3.8) and water quality standards at all sampling stations in the bay were achieved by 1973.

Kuruma prawn returned to the bay in 1983, and 527 species of organisms, including phytoplanktons, fish, and birds, have been found in the bay since 1989. The return of so many organisms to Dokai Bay and the recovery of biological communities, even if only a small step, will certainly encourage further environmental conservation efforts.

Kitakyushu City was awarded the "Global 500" prize from UNEP (the United Nations Environment Programme), as well as "UNCED Local Governors of Honors" from the United Nations Conference on Environment and Development in 1992.

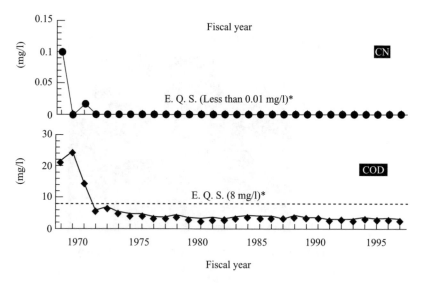

Figure 3.8 Trend of carbon, nitrogen, and COD concentrations at station D6 in Dokai Bay
Note: EQS = environmental quality standard.

ACKNOWLEDGEMENTS

The authors are indebted to the International EMECS Center for their invitation to the International Conferences on Environmental Management of Enclosed Coastal Seas, and to Masahiko Inatsugi of the EMECS Center for his kind revision of this chapter. Thanks are also due to Professor Kohichi Ohwada of the Ocean Research Institute of the University of Tokyo, who kindly gave the authors the chance to attend the International Conference on Man and Ocean organized by the United Nations University in Tokyo in October 1998.

REFERENCES

Okaichi, T. and T. Yanagi. 1997. Seto Inland Sea – Historical background. In: *Sustainable Development in the Seto Inland Sea, from the View Point of Fisheries*, T. Okaichi and T. Yanagi (eds), pp. 9–14. Terra Scientific Publishing, Tokyo.

Okaichi, T. 1997. Red tides in the Seto Inland Sea. In: *Sustainable Development in the Seto Inland Sea, from the View Point of Fisheries*, T. Okaichi and T. Yanagi (eds), pp. 251–304. Terra Scientific Publishing, Tokyo.

Vallejo, S.M. 1994. New structures for decision-making in integrated ocean policy. In: *Ocean Governance, Sustainable Development of the Seas*, P.B. Payoyo (ed.), pp. 71–95. United Nations University Press, Tokyo.

van Hoorn, H. 1989. Marine policies in ther North Sea region: The experience of the Netherlands. In: *Sea-use Planning and Coastal Area Management in Latin America and the Caribbean*, Report of the Expert Group Meeting, Economic Commission for Latin America and the Caribbean (ELAC), Santiago, Chile, 28th November– 1 December 1989.

Yamada, M., R. Takeuchi, S. Sueta, K. Kido, Y. Yabumoto, and Y. Yoshida. 1990. Recovery of aquatic animals in Dokai Bay, northern Kyushu, Japan. *Marine Pollution Bulletin*, 23: 201–207.

Yamada, M., T. Higashi, K. Hamada, K. Ueda, N. Eguchi, and M. Suzuki, 1998. Improvement and management of the environmental condition in the eutrophicated coastal bay with ecoremediation. *Environmental Science*, 11: 381–391 (in Japanese).

4

International marine science activities in Japan

Nobuhiko Handa

Introduction

As results of increased understanding of global biogeochemical cycles, particularly that of carbon, and of possible consequences of systemic and cumulative global changes, the IGBP (International Global Biosphere and Geosphere) was established in 1990, with a 10-year term. The JGOFS (Joint Global Ocean Flux Study), aimed at understanding the oceanic role in the uptake of the CO_2 produced from the burning of fossil fuels, is one of the IGBP's core projects. So far, Japanese JGOFS studies have been conducted in the western North Pacific, from the western equatorial area to the subarctic area, including the Bering Sea. Extensive ship observations of these areas have produced invaluable data, especially in relation to the carbon cycles. Some of the results obtained are as follows.

On the basis of data obtained in the area from the equator to 30°N at 137°E by the Meteorological Research Agency (MRA) Group, seasonal variation of pCO_2^s and long-term trends in the western North Pacific have been examined. The rate of increase of pCO_2^s in the subtropical areas was estimated at 1.8 $\mu atm/l/yr$, while the much lower value of 0.5 $\mu atm/l/yr$ was observed in the western equatorial areas. After detailed examination of the thermodynamic factors controlling the oceanic carbonate system in the western North Pacific, the increased rates of pCO_2^s in these oceanic areas were attributed to the ocean CO_2 uptake, which was equivalent to the TCO_2 increase at the rate of 1 $\mu mol/kg/yr$. The CO_2 outflux from the central and western equatorial Pacific was also examined and was found to decrease during El Niño, while an increase in La Niña was an indicator of intra- and inter-annual fluctuations of CO_2 outflux from the central and western equatorial Pacific.

Under the NOPACCS (North Pacific Carbon Cycle Study) research project, continuous survey of the fugacities of CO_2 in the surface seawater and air in the oceanic areas of 48°N through 15°S at 175°E was conducted from April to June (spring cruise) and August to October (summer cruise) every year from 1992 to 1997. Negative values of difference between pCO_2 in air ($pCO_2{}^a$) and surface sea-water ($pCO_2{}^s$){$pCO_2{}^s - pCO_2{}^a$} were observed in the restricted areas around 40°N whereas very high positive values, almost equivalent to air values, were found in the subtropical areas. The negative values, however, were common in whole areas of the ocean in spring time, except the area north of 45°N; further, the western North Pacific has a greater take-up of CO_2 in air in spring. $pCO_2{}^s$ tended to increase with seawater temperature in summer.

Extensive observation of the $pCO_2{}^a$ and $pCO_2{}^s$ was conducted in the northern North Pacific by a cargo ship cruising between Japanese ports and North American ports from March 1995 to the present. The data obtained by the National Institute of Environmental Science (NIES) are now being compiled and processed and a tentative rate of 0.29 GtC/yr for the oceanic uptake of CO_2 in this region has been established. The Japanese JGOFS committee has started to com-pile all the relevant data from the biogeochemical studies conducted in the west-ern North Pacific for the purpose of modelling the carbon cycle in the western North Pacific.

The Japanese GOOS (Global Ocean Observing System) was established in 1993 with a focus on the following:

• evaluation of oceanic transport of heat and materials in the North Pacific Ocean
• evaluation of fundamental elements in oceanic processes
• Design of an ocean observing system aided by high-resolution models of ocean circulation
• monitoring techniques producing time-series data on the oceanic environment
• Monitoring of ocean currents.

An international biomass symposium was held to review all the activities conducted by Japanese GOOS in the period 1993–1997. Positive and stimulating comments were offered by the reviewers in response to the newsletters published by Taira (1994–1998), and heat transport was determined in the western North Pacific off Japan, Izu Islands, and Alaskan stream. The reviewers also suggested that more international collaboration with the countries along the north-western Pacific Rim is desirable.

A satellite-based system for monitoring the ocean around Japan and an upper-ocean monitoring system in the mid-latitude North Pacific were developed to provide the oceanic science community with time-series SST and other variables. Positive comments and suggestions were forthcoming and international sharing of the data and modes of observation was urged.

Extensive modelling of Kuroshio and its related oceanic areas was carried out in order to understand the basic physical factors affecting large-scale circulation in the North Pacific. Response to the operation was positive and stimulating.

This project examined how ocean colour data could be assimilated into a coupled ecosystem model for estimating fluxes of particulate organic matter. This work will be developed for further understanding of the carbon cycle operating in the oceanic environment.

Background

The successful GEOSECS (Geochemical Sectional Studies) conducted in the 1970s opened a new era of global ocean observation through multidisciplinary cooperation in the fields of chemistry, biology, and physics. After GEOSECS, ocean communities throughout the 1970s and 1980s initiated various projects. However, emphasis in the geophysical community has shifted to the climate change issue as the possibility that the increasing amounts of greenhouse gases in the atmosphere might lead to global warming has become a general concern. After long discussion, the World Meteorological Organization (WMO) and the International Council of Scientific Unions (ICSU) agreed to submit collaborative worldwide observation data and scientific information collected through the World Climate Research Program (WCRP) and the International Geosphere-Biosphere Program (IGBP) to the Intergovernmental Panel on Climate Change (IPCC) to assist policy-makers in ensuring the sustainable development of human society. Towards this end, the IGBP established a Joint Global Ocean Flux Study (JGOFS) which was responsible for submitting ocean data and information to the IPCC.

Later, the Intergovernmental Oceanographic Commission (IOC) realized that the accumulation and circulation of physical and chemical data was an urgent issue if oceans were to be kept healthy and the supply of marine food resources was to be sustained. GOOS was launched in 1992 under the umbrella of the IOC, funded by the Japanese Ministry of Education, Science, Sport and Culture (MESSC), and involved the cooperative work of Japanese universities and national institutions in the 1990s. The JGOFS is to end by 2004 and therefore this programme is now in the synthesis and modelling phases; GOOS, however, will continue after 2004.

Review of scientific activities

With this scientific background, the present research status of the JGOFS and GOOS is reviewed.

JGOFS activities in Japan

As a result of increased understanding of the global biogeochemical cycle of carbon, and of the possible consequences of systematic and cumulative global changes, the IGBP was established in 1990 with a 10-year term. The JGOFS, aimed at understanding the oceanic role in the uptake of the CO_2 emitted by the burning of fossil fuels, is one of the IGBP's core projects. So far, nine Japanese research projects relevant to Japanese JGOFS activities have been conducted in the western North Pacific, spread over the western equatorial Pacific to the sub-arctic North Pacific, including the Bering Sea. Extensive ship observations of these oceanic areas have been obtained and invaluable data relevant especially to the carbon cycle have been accumulated.

JGOFS studies

Ocean Fluxes – Their Role in the Geosphere and Biosphere

Ocean Flux, PI: Hitoshi Sakai, Ocean Research Institute, University of Tokyo, 1991–1993

The major objectives of Ocean Flux were rather broad, but primarily focused on increasing understanding of the global biogeochemical cycling of biophilic elements such as carbon, nitrogen, phosphorus, and associated bioelements. The temporal changes of their concentrations through interactions across the major ocean boundaries such as land, atmosphere, and seafloor are another concern.

The ship observations, focused on the western North Pacific (40°S–44°N, 140°E–160°W), were conducted to determine the regional and seasonal variations of the fluxes of organic materials as well as biochemical elements such as carbon, nitrogen, phosphorus, and silicon throughout the air–sea boundary, ocean interior, and water–sediment boundary. These studies resulted in a great advance in understanding the long-range transport mechanism of terrestrial and anthropogenic materials from the land to the open ocean. This project also played an important role in understanding the dynamics of these biophilic elements in the North Pacific through to the South Pacific.

Most of the results and data obtained in this project were reported in a monograph (Sakai and Nozaki, 1995) as a part of the contribution of the Japanese JGOFS activities.

North Pacific Carbon Cycle Study

NOPACCS, PI: H. Tsubota, Hiroshima University, 1990–1996

NOPACCS focused primarily on carbon dynamics in the central North Pacific. The aim of the project was to conduct a continuous survey for the determination of the partial pressures of CO_2 in air ($PCO_2{}^a$) and surface seawater ($PCO_2{}^s$) in the oceanic area of 48°N through 15°S at 175°E from April to June (spring cruise) and August to October (summer cruise) from 1992 to 1997. The difference between PCO_2^s and PCO_2^a ($PCO_2^s - PCO_2^a$) was calculated. Negative values of the difference were observed in the subarctic areas north of 40°N, whereas positive values were determined in the subtropical areas, indicating that the subarctic and subtropical areas played important roles in the emission and absorption of CO_2 to the air overlaying the ocean surface. The area north of 45°N, however, revealed high absorption of CO_2 from air, especially in the spring, followed by the end of such an absorption process of CO_2 towards summer due to the increase in the sea surface temperature.

Sediment trap experiments were also conducted at four sites along 175°E from the equator to 48°N to determine the downward flux of organic carbon through the bottom of the euphotic layer. The annual flux of organic carbon to the under-lying water of the euphotic layer was almost equivalent to the dissolved oxygen consumption rate in the intermediate through deep waters. This indicates that biological agents dwelling there must degrade most of the organic matter supplied from the upper water layer to CO_2.

The NOPACCS group finally compiled all the chemical, biological, and phys-ical data to show the pattern of the carbon cycle in the central North Pacific (Tsubota et al., 1999).

Mass Flux Experiment in the Marginal Sea

MASFLEX, PI: K. Iseki, Western Regional Research Institute of Fisheries, 1992–1996

The continental shelves and slopes occupy less than 10 per cent of the surface areas of the world's oceans, yet more than 25 per cent of the ocean's total primary production originates there. Thus the JGOFS focused on this region in its study of the carbon dynamics of the marine environment. The aims of MASFLEX were the following: first, to understand the chemical, biological, and physical processes governing the fluxes and cycles of carbon, nitrogen, and phosphorus occurring in the continental margin of the East China Sea; second, extensive and intensive studies of the seasonal variations in primary productivity, downward flux of bio-genic carbon, and nitrogen; third, what amount of CO_2 exchanges between the air and the seawater; and fourth, what processes occur for lateral transport of car-bonic materials from the continental shelf area towards the deep water of the ocean interior.

The combined data showing the seasonal variability of primary production (Hama, 1994) and organic carbon flux (Iseki, Okamura, and Tsuchiya, 1994) indi-cated that far less than 5 per cent of organic carbon fixed by the primary production process was transported to the deep sea of the open ocean via the continental shelf and its slope. Tsunogai, Watanabe, and Sato (1999) conducted extensive studies determining total carbonate, pH, and alkalinity in the continental areas of the East China Sea and found that the partial pressure of CO_2 in the surface water (PCO_2^s)

of this area was much lower than that of air (PCO_2^a), indicating a distinctive trend in this oceanic area for strong absorption of atmospheric CO_2.

All the data obtained in this study have been compiled and processed for publication.

Monitoring Program for Air–Sea Exchange of Greenhouse Gases
in the Subarctic Region of the North Pacific

SKAUGRAN, PI: Yukihiro Nojiri, National Institute of Environmental Science, 1995–2001

The primary objectives of this study were to monitor the carbonate species, nutrients, phytoplankton pigments, and greenhouse gases in the northern North Pacific and to clarify the chemical, biological, and physical processes governing the air–sea exchange of CO_2 in the surface water. A cargo ship, *Skaugran*, makes 11 round trips per year between Japan and the west coast of North America in the subarctic area of the North Pacific. The eastbound cruise takes place between 30°N and 50°N in the subarctic North Pacific, while the great-circle track through the Bering Sea is used for the return cruise to Japan.

Extensive observation of PCO_2^a and PCO_2^s was conducted from March 1995 to 2000. The data obtained by the NIES are now being compiled and processed. Tentative results for CO_2 net flux calculation on the PCO_2^s grid function, sea surface wind field data of 1° resolution, and gas exchange coefficient after Wanninkhof (1992), showed 0.29 GtC/yr for the north of 37°N in the North Pacific, excluding marginal sea areas. These data were almost comparable to the values calculated by Takahashi *et al.* (1997), but the regional variability in the uptake of atmospheric CO_2 by surface water was distinctive. This phenomenon is primarily caused by the uptake of CO_2 due to the photosynthetic activities of phytoplankton. No close correlation between variability of PCO_2^s and Chl a was observed in this oceanic area, indicating that further intensive surveys of these two parameters should be undertaken.

Biogeochemical Studies of the North-western Subarctic Pacific

MIRAI Project, PI: M. Kusakabe, Japan Marine Science and Technology Center, 1997–2000

Because of strong turbulence in the sea surface in the northern North Pacific off Hokkaido, especially in wintertime, practically no oceanographic data have been available so far. This causes great difficulty in estimating the role of the northern North Pacific off Hokkaido in the carbon budget of the world's oceans. The *Mirai*, formerly a nuclear-powered test vessel, was inaugurated as an open oceangoing research vessel in 1998. The vessel is big enough to carry a hydrographic observation for biogeochemical and physical analyses even in oceanic environments with strong sea surface turbulence.

The main aims of this project are to collect hydrographic data such as seawater temperature, salinity, nutrients, carbonate system components, and current direction and velocity, and to determine the regional and seasonal variability of biogeochemical parameters such as primary productivity and downward flux of organic matter, as well as the standing stock of biological materials. The current goal of the project is to compile the chemical, biological, and physical data to calculate the carbon budget of this ocean area.

In October–December 1998 and May 1999 the *Mirai* cruised the area off Hokkaido spanning 35–51°N, 145–165°E. The data are now being processed.

Time-Series Observations at the Fixed Station, KNOT,
in the Western Subarctic Pacific

KNOT, PI: Yukihiro Nojiri, National Institute of Environmental Science, 1997–2000

The oceanographic community has recently realized the importance of long-term observation of the marine environment at strictly selected experiment sites in understanding general trends of spatial and seasonal variation of marine environmental parameters related to the carbon cycle of the ocean. Thus far, two sites have been selected in the central North Pacific off Hawaii (HOT) and in the central Atlantic off Bermuda (BAT). The BAT study has a long history of over 35 years, while data from HOT have been accumulated since 1988. At both sites, seasonal variation in total inorganic carbon concentration definitely occurred, and it was observed that atmospheric CO_2 influences seasonally and to a large degree the total carbonate material concentration without any significant time delay.

The Japanese JGOFS group, realizing that it was important to conduct long-term observations of marine environmental parameters in a site within the western subarctic gyre, commenced a time-series observation in June 1998 at a site fixed at 44°N, 155°E (KNOT). Research vessels from various institutions conducting basic hydrographic cast and primary productivity measurements visited the site 13 times in 1998. The only sediment trap experiment was carried out by R/V *Mirai*.

Some results of the study have been disclosed, indicating that marked seasonal variability in chlorophyll occurs with maxima in early June and mid-October corresponding to the spring and autumnal blooming of phytoplankton, characteristic of the western subarctic North Pacific; and total inorganic carbon showed marked seasonal variation in 1998 with a minimum in August, increasing towards December, and reaching a maximum some time in the following winter. Organic carbon flux and primary productivity data are not yet available at this time.

International Cooperative Studies on the Advanced Technology of
Global Mapping of Biogeochemical Parameters

Global Mapping, PI: Hotaka Kawabata, Japan Geological Survey, 1998–2002

This was planned as a joint project between the ocean and terrestrial ecosystem communities with the aim of estimating how much anthropogenic CO_2 is fixed by the ocean and terrestrial ecosystem on the basis of data obtained by both field observation and satellite image analyses. Oceanic works determining carbonate materials are an important part of this study to assess the amount of CO_2 that is fixed or released through the air–sea interface in relation to ENSO (El Niño Southern Oscillation).

Results obtained so far include the following. Seasonal variation and long-term trends in PCO_2^s in the western North Pacific were examined on the basis of data obtained in the area from the equator to 30°N along 137°E by the MRI group. Findings show an increasing rate of 1.8 µatm/yr for PCO_2^s in the subtropical areas. A much lower value of 0.5 µatm/yr, however, was observed in the western equatorial areas. From a detailed examination of the thermodynamic factors controlling the oceanic carbonate system in the western North Pacific, it was determined that the increasing rates of PCO_2^s in these oceanic areas were due to an ocean CO_2 uptake that was equivalent to the TCO_2 increase at the rate of 2 µmol/kg/yr. The CO_2 outflux from the central and western equatorial Pacific was also examined and it was found that the outflux tended to decrease during the El Niño event, and to increase in the La Niña event, suggesting intra- and interannual fluctuations of CO_2 outflux from the central and western equatorial Pacific, including the subtropical South Pacific.

Further extensive studies of carbonate material in the equatorial and subtropical oceanic areas will be carried out under this project. Japanese JGOFS are eagerly awaiting these data in order to combine them with the carbonate data obtained in the arctic and subarctic areas of the North Pacific, in an effort to understand the activity of anthropogenic CO_2 on a global scale.

GOOS activity in Japan

The international community has recently realized that the ocean may hold important clues to solving the problems of not only global warming but also food supply. Therefore, the IOC decided to establish the GOOS in cooperation with the WMO, the United Nations Environment Programme (UNEP), and the ICSU in 1992. The aims of this programme are monitoring and prediction of the global climate and ocean changes, and collection, analysis, and distribution of oceanic data for the maintenance of ocean health, management of marine biological resources, and for global environmental studies.

Japan's MESSC decided to start this programme in Japan in 1993 with the participation of research scientists from various universities and national institutes.

The Japanese GOOS focused on several primary subjects (Taira, 1994, 1995, 1996, 1997, 1998).

Evaluation of oceanic transport of heat and materials in the North Pacific Ocean

Heat and volume transportation by ocean circulation were observed by means of moored current meters and inverted echo-sounders, CTD casts, acoustic drop-sondes, and satellite altimeters in two sections across the Kuroshio Current in the western North Pacific, off Shikoku and on the Izu Ridge. Another section was also accepted to determine the heat and volume transportation in the subarctic gyre around the Date Line. In June 1993, four multipath-inverted echo-sounders were moored over the Izu Ridge to detect volume transport and the path of the Kuroshio. In October 1993, 24 current meters and nine inverted echo-sounders were moored at nine stations off Shikoku. Turbulent processes, which affected the ocean currents and heat flux, were studied using a towed thermister chain.

Evaluation of fundamental elements of ocean processes

Fluxes of sensible and latent heat and momentum across the ocean surface are estimated by satellite data, and validated by data from ships and buoys. A surface buoy which can withstand rough seas was developed and used to monitor surface fluxes. Surface fluxes determined in the close coupling between ocean and atmosphere were identified as fundamental processes driving ocean circulation.

Design of ocean observing system aided by high-resolution models
of ocean circulation

Numerical models of general and regional circulation have been developed to identify the key elements and locations for monitoring. The models are essential for the interpolation of data, because ocean observations with uniform spatial and temporal scales over widely disparate areas are impractical, and for forecasting through the analysis of oceanic states in the past.

Monitoring techniques securing time-series data on the ocean environment

An efficient and reliable technique of analysing dissolved gases and radioactive nuclei was developed to monitor the budget of greenhouse gases and deep circulation. Field observations were made at selected stations with a time interval sufficient to monitor changes in the oceanic environment.

Monitoring of ocean currents and biomass abundance using new techniques

The study of biological activities and environments is essential to the understanding of the material cycles in the ocean. Acoustic techniques and an algorithm for the processing of satellite data have been developed to evaluate plankton density and biological environments. Current fields are monitored by acoustic Doppler current profilers (ADCPs) and induced voltage of submarine cable across a strait.

Conclusion

The UNESCO/IOC/WESTPAC Fourth International Scientific Symposium was held for an international review of all Japanese GOOS activities conducted in the period 1993–1997. The reviews commented positively on the volume and heat transport studies conducted in the south of Japan, the Izu Islands, and Alaskan stream, but suggested that more international collaboration with countries along the north-western Pacific rim would be desirable.

A satellite-based system for monitoring the ocean around Japan and an upper ocean monitoring system in the mid-latitude North Pacific were developed to provide time-series SST and other variability data to the ocean community. These received positive comments; international sharing of the data and modes of water observation was suggested.

Extensive modelling studies on the Kuroshio and its related oceanic areas were developed to give an understanding of the basic physical factors affecting large-scale circulation in the North Pacific. Intentional cooperation in this study was encouraged.

The project also examined how ocean colour data can be assimilated into a coupled ecosystem model for the estimation of fluxes of particulate organic matter. This work is expected to develop, leading to further understanding of the carbon cycle operating in the oceanic environment.

REFERENCES

Hama, T. 1997. Seasonal change of the primary productivity in the East China Sea. In: *Global Fluxes of Carbon and Its related Substances in the Coastal Sea–Ocean–Atmosphere System,* S. Tsunogai (ed.), Proceedings of 1994 Sapporo IGBP Symposium, pp. 74–79.

Iseki, K., K. Okamura, and Y. Tsuchiya. 1997. Seasonal variability in particle distribution in the East China Sea. In: *Global Fluxes of Carbon and Its related Substances in the Coastal Sea–Ocean–Atmosphere System,* S. Tsunogai (ed.), Proceedings of 1994 Sapporo IGBP Symposium, pp. 189–197.

Sakai, H. and Y. Nozaki. 1995. *Biogeochemical Process and Ocean Flux in the Western Pacific,* 672pp. Terra Scientific Publishing, Tokyo.

Taira, K. 1994, 1995, 1996, 1997, 1998. International Cooperative Research Program on Global Ocean Observing System. Sponsored by Ministry of Education, Science and Culture, Japan. *Newsletter* No. 1 (1994), No. 2 (1995), No. 3 (1996), No. 4 (1997) and No. 5 (1998), Ocean Research Institute, University of Tokyo.

Takahashi, T., R.A. Freely, R. Weiss, R.H. Wanninkof, D.W. Chipman, S.C. Sutherland, and T. T. Takahashi. 1997. Global air–sea flux of CO_2: An estimate based on measurement of sea–air PCO_2 difference. *Proceedings of the National Academy of Sciences, USA,* No. 94, pp. 8292–8299.

Tsubota, H., K. Harada, K. Ishida, J. Ishizaka, and Y. Watanabe. 1999. Hydrographic data of NOPACCS. In: *Secind International Symposium on CO_2 in the Ocean (Extended Abstract),* p. 29, National Institute for Environmental Studies.

Tsunogai, S., S. Watanabe, and T. Sato. 1999. Is here a continental shelf pump for the absorption of atmospheric carbon dioxide. *Tellus,* 51B: 701–712.

Wanninkhof, R. 1992. Relationship between wind speed and gas exchange over the ocean. *Journal of Geophysical Res*earch, 97: 773–782.

5

The UNU's international marine environment research networks: An approach towards sustainable seas in the twenty-first century

Juha I. Uitto and Zafar Adeel

Introduction

The United Nations University (UNU) has played a key role in developing networks of researchers and in initiating global and regional research projects related to the global environment. The research undertaken by the UNU relies on collaborative efforts to deal with complex global problems, such as those related to the marine environment. The success of the network development approach has provided an added initiative for continuing activities over the long term. A good example of this is a UNU programme that focuses on environmental monitoring in the coastal hydrosphere in the East Asian region (Adeel, 1999). "Coastal hydrosphere" refers to freshwater and seawater resources in the coastal areas as well as the living resources within these waters. The primary objectives of this programme are to build the capacity for coastal monitoring, develop a comprehensive database of pollution in coastal areas, and develop integrated guidelines for coastal development in East Asia. It is the authors' belief that an open and scientifically based discussion of coastal resource management within this network can lead to enlightenment of participants and government policy-makers alike.

There exists a complex relationship between human activities and the marine and coastal environments. There are a number of human activities and economic ventures based on the resources these environments provide, and a vast majority of these anthropogenic activities impose a burden on these very same resources.

On the whole, oceans and seas have numerous functions that are essential to both ecosystem health and human society. The oceans play a central role in regulating climate. More specifically, oceans and atmosphere are intimately coupled with the global hydrological cycle. The cooling or heating of the sea surface directly controls the atmospheric characteristics of overlying air masses (Schotterer and Andermatt, 1990). This, in turn, plays a key role in defining global weather, as well as long-term climate patterns. In additions, coral reefs are recognized as important sinks of carbon dioxide, and so play a significant part in climate patterns.

Throughout history, oceans and seas have been essential for transportation, and have brought distant peoples and cultures in closer contact. They provide a source of food for a large segment of human society and they are increasingly a source of recreation for people. Fish account for about 20 per cent of the protein in the human diet and about 80 million tonnes are consumed each year (WRI, 1998). Fishing and related industries are an important factor in global employment. According to one estimate, marine-related industries contribute over seven trillion US dollars to the global economy (Borgese, 1998).

Until recently, the oceans were considered inexhaustible sinks where waste and other pollutants could be dumped safely. It is now recognized, however, that marine and coastal environments are strongly affected by human activities, and that substances released into oceans can degrade the chemical composition of seawater. Similarly, overfishing has resulted in the increasing depletion of fishing stocks. The global fish catches have peaked and many larger high-value species have become endangered. According to the United Nations Food and Agriculture Organization, all of the world's major fishing regions are harvested at or beyond their capacity (FAO, 2001).

The long-term implications of depleted marine resources and deteriorating environmental conditions are now becoming increasingly obvious. These include threats identified as an indirect result of global environmental change. Global warming is thought to have raised the temperature of the oceans around the equator to a level endangering the sensitive coral reefs. Depletion of the atmospheric ozone layer is causing more ultraviolet radiation to reach the earth's surface, affecting the reproductive mechanisms of plankton. These processes are still little understood, however, making rational decision-making difficult. There is a need to increase further the scientific knowledge that could provide a basis for such decision-making.

The oceans should be seen as a common resource belonging to everybody, and everyone is equally responsible for their wise and sustainable use. Solid science and innovative technologies can provide a basis for sound policy-making for sustainable development of coastal and marine areas. Similarly, capacity-building and education are needed to enable countries and the various sectors to tackle environmental problems (Uitto, 2003). This holistic approach towards the oceans and their resources is a keystone of the activities undertaken at the UNU. This chapter describes the ongoing and future activities for coastal and marine areas, while providing a background and justification for them.

Background

The earliest concerns about marine pollution, raised in the 1950s, related to the release of artificial radionuclides from nuclear power plants into the oceans and the atmosphere (Goldberg, 1991). Since marine food sources and water were affected, these radionuclides were perceived as a potential public health hazard. In the 1960s the environmental movement became more active, and an increasing number of studies focused on the issue of marine pollution. Pesticides, like DDT and other halogenated hydrocarbons, became well known for their accumulation in the marine food chain, resulting in frequent nesting failure for fish-eating seabirds (Fry, 1999).

Coastal zones the world over, approximately 40,000 km in length, are particularly vulnerable to human influences. These are the areas where urban growth is most rapid, resulting in excessive stress on the natural resources (Goldberg, 1991). Encroachment of human activities by way of building, changing land use, tourism, and pollution from both industry and agriculture strains the natural coastal ecosystems. In particular, for developing countries with relatively high population growth and limited infrastructure, financial, and manpower resources, such developments are rapid and difficult to manage.

Although land-based sources of pollution pose the most important threat to coastal and marine environments, maritime accidents have also made a very significant contribution to their deterioration. Large-scale oil spills like the 147,000-litre *Exxon Valdez* spill in Alaska were commonplace in the late twentieth century (Kelso and Kendziorek, 1991; Davis, 1996). These spills kill the fish, marine mammals, and seabirds that live in coastal habitats (e.g. 1,000 sea otters and about 36,000 seabirds in the case of the *Exxon Valdez* spill; Maki, 1991). The extent of damage caused by such maritime accidents can be estimated by observing the overall coastal deterioration resulting from the *Exxon Valdez* accident in Alaska, as shown in Table 5.1.

Unfortunately, Japan has also suffered its share of oil spills in its territorial waters in the past years. The most devastating one was in January 1997, when the Russian tanker *Nakhodka* broke up in the Sea of Japan and released its cargo of fuel oil into the sea. The oil polluted the coast of nine prefectures and the spill was one of the worst marine pollution disasters in Japan (Kobayashi, 1998).

Table 5.1 The impact of the *Exxon Valdez* oil spill in Alaska, USA

Area	Total coastline (km)	Oil-impacted coastline (km)	
		May 1989	July 1990
Prince William Sound	4,926	575	148
Gulf of Alaska	10,210	1,178	39
Total	15,136	1,753	187

Source: Maki, 1991

Human activities have also destroyed a number of coral reef ecosystems (Borgese, 1998). The coral reefs of the world are delicate ecosystems that play an important role as carbon dioxide sinks, and provide a habitat for a wide variety of fish and marine fauna. High pollution levels and increasing water temperatures can lead to their bleaching and ultimate death. This effect is also indirectly linked to global warming and stripping of the ozone layer. Due to various anthropogenic threats, current estimates place between 30 and 40 per cent of the world's reef area at medium to high risk (Bryant *et al.*, 1998). For example, in South-East Asia it is estimated that 10 percent of the reefs have been destroyed and in some places, like in Indonesia, most of the reefs have been severely damaged (Hidayati, 2000).

In this context, it is crucial that we manage our marine resources globally and at an international level. In order to meet the challenges of sustainable use and development of global marine resources, close international collaboration is required. Another equally important factor is that the problems faced in the marine environment today are of a multidisciplinary nature. These problems can no longer be considered the sole domain of natural sciences. Rather, they require input from experts in various fields, such as politics, social sciences, economics, history, and geography. Science must provide the knowledge and information that is needed for informed decision-making. At the same time, it is important to realize that politicians, bureaucrats, large enterprises, and society as a whole base decisions on political, economic, strategic, and other priorities. Therefore, it is essential that the international scientific community shares information and research results across various disciplines and geographic boundaries. A very practical way of achieving this is to develop and promote international and multidisciplinary networks of researchers.

The UNU's involvement in marine and coastal affairs

The mandate of the UNU fits in with the general approach of network development. The UNU is enjoined to carry out pure and applied research, advanced training and capacity-building, and dissemination of knowledge on pressing global issues. The university seeks innovative solutions to these complex global problems and serves as a think-tank to the United Nations. As an autonomous academic organization under the umbrella of the United Nations, it is well placed to mobilize intellectual resources while remaining free from political pressures. The university also pays particular attention to capacity-building and dissemination of knowledge in the developing countries, which face multiple development and management challenges.

The work of the UNU is divided into two broad categories. The first deals with global issues related to the environment and sustainable development. The second is related to issues regarding peace and governance. In view of the holistic perspective needed for the sustainable management of marine environments, both of these categories are relevant here. Similarly, equal attention is paid to natural and social sciences, as well as to governance and legal frameworks.

The primary *modus operandi* adopted at the UNU for achieving its objectives revolve around development of networks. Large numbers of scholars, educators, and students participate in various research, training, and capacity-building activities. In particular, the UNU places emphasis on working together with scholars and institutions in the developing countries. The objective is to develop science-based policy, and to contribute to indigenous capacity-building.

The UNU's priorities and actions in the field of environment and sustainable development in general, and coastal and marine issues specifically, are guided by the implementation plan of the World Summit on Sustainable Development (WSSD 2002) held in Johannesburg, South Africa. These also relate to the Agenda 21, which provides a blueprint for the international community to chart its way towards the sustainable development of marine and coastal resources. The Agenda 21 chapter on "Protection of oceans, all kinds of seas, including enclosed and semi-enclosed seas, and coastal areas and the protection, rational use and development of their living resources" starts with the following statement:

The marine environment – including the oceans and all seas and adjacent coastal areas – forms an integrated whole that is an essential component of the global life-support system and a positive asset that presents opportunities for sustainable development. (United Nations, 1992: 130)

Agenda 21 also identifies the difficulties and obstacles in achieving this integrated approach. For example, it pinpoints the potential problems in the management of high-seas fisheries, including the adoption, monitoring, and enforcement of effective conservation measures. Fishing fleets operating in international waters are utilizing inappropriate and indiscriminate fishing methods, resulting in over-harvesting of living marine resources and frequently killing species, such as dolphins, that are not targeted. Similarly, monitoring and preventing marine pollution on the high seas, including dumping of hazardous wastes, are very difficult. Furthermore, in many cases different international agreements and conventions come into conflict. A case in point would be the conflict between trade and environment. The World Trade Organization (WTO) has ruled that banning imports of fish caught by environmentally detrimental methods would constitute illegal trade restrictions. There is a need to solve these contradictions in international agreements and governance. Independent international research, such as that carried out under UNU auspices, can provide invaluable assistance in solving these problems.

The UNU has an extended history in promoting ocean governance with a view to achieving sustainable utilization of marine space in international waters or waters that fall within the jurisdiction of more than one state. In this context, two early conferences are of particular note. First, the International Conference on the Sea of Japan: Transnational Ocean Resource Management Issues and Options for Cooperation was co-organized by the UNU, the East-West Center, and other partners in Niigata, Japan, in 1988. This conference focused on issues pertaining to marine resources and their use, and explored possibilities for regional cooperation in their management in Northeast Asia. Second, the International Conference on East Asian Seas: Cooperative Solutions to Transnational Issues was co-organized by the UNU, the East-West Center, the Korea Ocean Research and Development Institute (KORDI), and other partners in Seoul, Korea, in 1992. Issues here ranged from fisheries to petroleum and minerals, trade and shipping, and environmental pollution.

As an outcome of these conferences, it was recognized that the Sea of Japan was already suffering from pollution, especially mercury contamination. The great potential for tourism in marine and coastal areas requires that marine pollution in these regions be effectively controlled. Emphasis was placed on the need for heightened environmental awareness in the countries in this region and for incorporating environmental costs into economic calculations concerning the utilization of the sea (Valencia, 1989). The meetings also outlined specific cooperative mechanisms and discussed their design and implementation. The UNU has also participated in the *Pacem in Maribus* process to develop new mechanisms for ocean governance (Payoyo, 1994).

In addition to the focus on ocean governance, the UNU has adopted a geographical approach, focusing on critical ecosystems and regions. This includes coastal zone ecosystems as a high-priority area. The UNU's work on coastal resources management stems from the early period of the university's existence. The networking approach has taken into account the multidisciplinary nature of these problems as well as the regional outlook of the strategies for solving them. Several examples of the UNU's efforts in the early period after its inception, highlight this overall approach. As an example, a project on "Coastal Resources Management" related to the humid tropical coasts in Indonesia was initiated in 1979. The project focused on the Cimanuk Delta in northern Java, a rapidly changing deltaic system, typical of intensively used deltas in Indonesia (Bird and Ongkosongo, 1980; Bird and Soegiarto, 1980). The project concentrated on the organization and supervision of training and research courses and management workshops on selected coastal environments, particularly in tropical countries.

Apart from damaging coastal ecosystems, pollution has been recognized as causing concrete and severe health problems in communities dependent on marine products for their livelihood. Perhaps the most famous case was that of industrial methyl-mercury poisoning in Minamata on the south-western Japanese island of Kyushu. Although the Minamata disease was identified as early as 1956 when patients who had consumed seafood caught in Minamata Bay developed severe neurological disorders, frequently leading to death, its social and political repercussions have continued until today (Carpenter, 1992; Maruyama, 1996; Ui, 1992). The UNU has been studying community responses to and the long-term impacts of industrial disasters (Mitchell, 1996). This type of work, looking at the community recovery from extreme disruptions caused by environmental pollution incidents from a social point of view, has been generally neglected. The UNU recognizes that investigating such social aspects is important in assisting communities struggling to return to their "normal states".

The UNU was also involved with the Asia-Pacific Mussel Watch (APMW), which was undertaken as an integral part of the International Mussel Watch Programme. That programme aimed to provide an assessment of the status and trends of chemical contaminants in the world's coastal waters, utilizing mussels and other sentinel bivalves as indicators (Sudaryanto et al., 2002). The targeted pollutants of that study were butyltin compounds and organochlorine compounds (such as PCBs, DDT, and others). These pollutants are known to have caused adverse hormonal effects in various species of mussels and are thus considered to be endocrine disruptors. In this project, the UNU joined hands with the UNESCO Intergovernmental Oceanographic Commission (IOC) and the United Nations Environment Programme (UNEP). The UNU was charged with the coordinating role for the APMW, while Ehime University in Japan took a lead role in monitoring coastal environmental quality in the Asia-Pacific region. At the same time, the Environmental Research and Training Centre (ERTC, Thailand), in collaboration with the UNU, took the lead in capacity-building through training workshops and inter-calibration exercises.

In another initiative, the UNU developed a project on research and conservation of mangroves in the Asia-Pacific region jointly with the International Society for Mangrove Ecosystems and the UNESCO Man and the Biosphere programme. This initiative builds upon the university's earlier work (Kunstadter, Bird, and Sabhasri, 1986). Mangroves are important and complicated ecosystems that are common in the subtropical and tropical parts of the Asia-Pacific region (Vanucci, 2004). They are irreplaceable in the ecosystem as spawning grounds for fish and shellfish, a habitat for birds and other plants and animals, and as part of the hydrological balance. The ecological, environmental, and socio-economic importance of mangroves is widely accepted.

Mangroves are vulnerable to sudden or drastic changes in the environment, and they do not recover spontaneously after the impact of natural or man-made catastrophic events such as cyclones or total felling of the mangrove forests. They are seriously affected by pollution, as well as by sedimentation caused by deforestation on land. There is an urgent need for bringing mangroves into the forefront of habitat protection. In this context, the UNU has teamed up with UNESCO and ISME to undertake a three-year research and training project: Asia-Pacific cooperation for the sustainable use of renewable natural resources in biosphere reserves and similar managed areas (ASPACO).

The UNU's coastal and marine programmes into the new millennium

The success of the network development approach has provided the added incentive for continuing activities over the long term. In this regard there is one programme that has been active for several years and will continue well past the threshold of the new millennium: "Environmental Monitoring and Governance in the East Asian Coastal Hydrosphere". In terms of implementation, the project comprises three parallel and mutually complementary components: coastal hydrosphere monitoring (both seawater and freshwater) for endocrine disruptor compounds (EDCs) and the development of a database of monitoring information; development of an international expert network; and guidelines for the development of consistent coastal management programmes in East Asia (Adeel, 1999). Nine countries are participating in the first component of this project: China, Indonesia, Japan, the Republic of Korea, Malaysia, the Philippines, Singapore, Thailand, and Viet Nam.

The primary objectives of the project are as follows.

- Capacity-building for coastal hydrosphere monitoring: learning the latest techniques for monitoring of EDCs in freshwater and seawater will further the capacity-building component of the project.
- Development of a regional monitoring database: the database will contain the data collected for the presence of land-based pollutants, particularly suspected EDCs, in the coastal hydrosphere samples. This database is anticipated to serve as the backbone for implementing coastal management policies in the participant countries.
- Development of guidelines for consistent coastal management plans in East Asia: most participant countries have a similar coastal environment and face similar challenges in managing coastal resources. Therefore, coastal management programmes and practices in these countries can be made mutually consistent and comparable.

Coastal hydrosphere monitoring means monitoring pollution in both freshwater input to the coastal areas and the seawater. The focus of this monitoring is on compounds that are suspected to be EDCs. More specifically, the targeted compounds include pesticides and organochlorine compounds that result from land-based coastal pollution. Recent research has revealed that several animal species have suffered adverse health effects from exposure to chemical pollutants that are suspected to interact with the endocrine system (Kavlock et al., 1996). Several researchers have reported endocrine-mediated anomalies in specific populations of invertebrate, fish, avian, reptilian, and mammalian species. Most commonly, these health effects are a result of relatively high-level exposure to organochlorine compounds (such as DDT and its metabolites, polychlorinated biphenyls (PCBs), and dioxins) and some naturally occurring oestrogens. Currently, about 50 chemical compounds are strongly suspected to disrupt the hormonal systems of affected animal life and humans. Even more compounds (up to 15,000 chemicals!) are currently under scrutiny as possible EDCs. The effects of these contaminants on humans, either through direct ingestion or through contaminated sources of food such as fish, are currently under intense research.

Reliable regional data in East Asia are difficult to access. The UNU project aims to develop a regional database while facilitating transfer of technology and knowledge to the participating laboratories. Technology transfer and capacity-building is, in part, done through specialized training and hands-on experience. At the same time, the researchers involved in this project work towards enhanced regional cooperation and promote exchange of data and monitoring results. The final product of this project component is a comprehensive database of targeted pollutant levels in coastal areas. The database is compiled into and disseminated from LandBase (landbase.hq.unu.edu), which is a web-based interface for data and information presentation.

The information contained in LandBase is used in a parallel programme to develop guidelines for a consistent regional coastal management programme. A summary of policy recommendations and guidelines, based on the work under-taken within the project, has recently been published (Adeel and King, 2002). Two workshops have also been organized to develop regional guidelines for water quality: "Environmental Quality Guidelines and Capacity Development" (Gwangju, Republic of Korea, 26–27 January 2003) and "Capacity Development for Monitoring of POPs in the East Asian Hydrosphere" (Tokyo and Hadano, Japan, 1–2 September 2003).

Conclusions

Various aspects of coastal resource management are important in developing coherent and consistent policies and are addressed in this project. First, it is important to preserve, protect, restore, and enhance the natural coastal resources. This includes identifying species and processes at risk and potential land-based sources. In this context, preliminary ecological risk assessment can estimate potentially adverse effects. If necessary, restoration technologies must be applied to the affected areas. Second, it is essential to assist governments in effectively exercising their responsibilities to make appropriate use of land and water resources in the coastal zone, as perceived under UNEP-GPA. In this regard, full consideration has to be given to ecological, cultural, historic, and aesthetic values. Appropriate emphasis also must be given to sustainable eco-nomic development along coastal areas. Finally, participation, cooperation, and coordination by the public, governmental agencies, and non-governmental organizations (NGOs) should be strongly encouraged. These key stakeholders have a significant role to play in the development and implementation of coastal management plans.

Open and scientifically based discussion on these aspects of coastal resource management can improve the level of awareness of participants and government policy-makers alike. It is anticipated that the guidelines and recommendations from UNU programmes will be used by policy-makers in developing coastal management plans and programmes. Therefore, effectively disseminating relevant results to key stakeholders within each participant country is essential.

The collaborative efforts undertaken by the UNU and its partners bring together researchers from vastly diverse geographical and educational backgrounds. It is hoped that the sustainable use of marine and coastal resources can be promoted through targeted research, capacity-building, and knowledge transfer as a result of this network development.

REFERENCES

Adeel, Z. 1999. Key environmental issues related to the coastal hydrosphere in East Asia: Directions for future work. In: *Proceedings of Environmental Governance and Analytical Techniques: Environmental Issues Related to EDC Pollution in East Asia*, UNU, Tokyo, Japan, 9–10 February 1999, United Nations University, Tokyo, pp. 82–86.

Adeel, Z. and King, C. 2002. *Conserving Our Coastal Environment*. United Nations University, Tokyo, Japan.

Bird, E.C.F. and O.S.R. Ongkosongo. 1980. *Environmental Changes on the Coasts of Indonesia*. United Nations University Press, Tokyo.

Bird, E.C.F. and A. Soegiarto (eds). 1980. *Proceedings of the Jakarta Workshop on Coastal Resources Management*. United Nations University Press, Tokyo.

Borgese, E.M. 1998. *The Oceanic Circle – Governing the Seas as a Global Resource*. United Nations University Press, Tokyo.

Bryant, D., L. Burke, J.W. McManus, and M. Spalding. 1998. *Reefs at Risk: A Map-Based Indicator of Potential Threats to the World's Coral Reefs*. World Resources Institute, Washington, DC.

Carpenter, R.A. 1992. *Industry, The Environment and Human Health: In Search of a Harmonious Relationship*. Report of the conference held in Minamata, Japan, 14–15 November 1991. United Nations University Press, Tokyo.

Davis, Y.D. 1996. The *Exxon Valdez* oil spill, Alaska. In: *The Long Road to Recovery: Community Responses to Industrial Disaster*, J.K. Mitchell (ed.). United Nations University Press, Tokyo.

FAO. 2001. *World Fisheries and Aquaculture Atlas*: Food and Agriculture Organization.

Fry, M. 1999. Endocrine disrupting chemical impacts on coastal birds. In: *Proceedings of Environmental Governance and Analytical Techniques: Environmental Issues Related to EDC Pollution in East Asia*. UNU, Japan, 9–10 February 1999. United Nations University, Tokyo, pp. 167–203.

Goldberg, E.D. 1991. *Coastal Zone Space: Sites for Conflict*. UN University Lectures 3, United Nations University, Tokyo.

Hidayati, D. 2000. *Coastal Management in ASEAN Countries: The Struggle to Achieve Sustainable Coastal Development*, United Nations University, Tokyo.

Kavlock, R.J., G.P. Daston, C. DeRosa, P. Fenner-Crisp, L.E. Gray, S. Kaattari, G. Lucier, M. Luster, M.J. Mac, C. Maczka, R. Miller, J. Moore, R. Rolland, G. Scott, D.M. Sheehan, T. Sinks, and H.A. Tilson. 1996. Research needs for the risk assessment of health and environmental effects of endocrine disruptors: A report of the U.S. EPA-sponsored workshop. *Environmental Health Perspectives*, No. 104, Supplement 4.

Kelso, D.D. and M. Kendziorek. 1991. Alaska's response to the *Exxon Valdez* oil spill. *Environmental Science and Technology*, 25(1): 16–23.

Kobayashi, Y. 1998. Management of ocean disasters. *Work in Progress*. United Nations University Press, Tokyo.

Kunstadter, P., E.C.F. Bird, and S. Sabhasri (eds). 1986. *Man in the Mangroves: The Socio-Economic Situation of Human Settlements in Mangrove Forests*. United Nations University Press, Tokyo.

Maki, 1991. The *Exxon Valdez* oil spill: Initial environmental impact assessment. *Environmental Science and Technology*, 25(1): 24–29.

Maruyama, S. 1996. Responses to minamata disease. In: *The Long Road to Recovery: Community Responses to Industrial Disaster*, J.K. Mitchell (ed.), pp. 41–59. United Nations University Press, Tokyo.

Mitchell, J.K. (ed.). 1996. *The Long Road to Recovery: Community Responses to Industrial Disasters*. United Nations University Press, Tokyo.

Payoyo, P.B. (ed.). 1994. *Ocean Governance: Sustainable Development of the Seas*. United Nations University Press, Tokyo.

Schotterer, U. and P. Andermatt. 1990. *Climate – Our Future*. Kümmerly and Frey Geographical Publishers, Berne.

Sudaryanto, A., S. Takahashi, I. Monirith, A. Ismail, M. Muchtar, J. Zheng, B.J. Richardson, A. Subramanian, M. Prudente, D.H. Nguyen, and S. Tanabe. 2002. Asia-Pacific mussel watch: Monitoring of butyltin contamination in coastal waters of Asian developing countries. *Environmental Toxicology and Chemistry*, 21: 2119–2130.

Ui, J. 1992. Minamata disease. In: *Industrial Pollution in Japan*, J. Ui (ed.), pp. 103–132. United Nations University Press, Tokyo.

Uitto, J.I. 2003. International Cooperation for Sustainable Coastal and Marine Management in East Asia. *Geographical Review of Japan*, 76(12): 869–881.

United Nations. 1992. *Agenda 21*. United Nations, New York.

Valencia, M.J. 1989. *International Conference on the Sea of Japan*. Occasional Papers of the East-West Environment and Policy Institute, No. 10, East-West Center, Honolulu, HI.

Vanucci, M., (ed.) 2004. *Mangrove Management and Conservation: Present and Future*. United Nations University Press, Tokyo.

World Resources Institute (WRI). 1998. *World Resources 1998–99*, WRI, Washington, DC.

Part II

Case studies of marine pollution in the world

6

Environmental problems in the coastal waters of China

Zhou Kaiya

Introduction

In 1998, China's marine environmental problems drew frequent attention in the press. In April many newspapers carried stories about the red tide outbreaks in the Pearl River estuary and off the coast of Hong Kong. The State Oceanic Administration's June declaration that "The Bohai Sea will soon become a 'dead sea' if pollution continues" appeared on the *Beijing Youth Daily*'s front page (Yuan, 1998). In September, a report to the Standing Committee of the National People's Congress on marine environment protection by Zou Jiahua, the committee's vice president, was outlined in the *Xinhua Daily*. In the report, an earnest Zou pointed out that the water quality of China's coasts had been decreasing with each passing year (Zou, 1998). These press reports focused public attention on the problems facing the marine environment.

China has a mainland coastline of more than 18,000 km, an island coastline of more than 14,000 km, a vast continental shelf, and an exclusive economic zone (EEZ) (Figure 6.1). In the 1980s, the State Oceanic Administration established both the All Area Ocean Monitoring and the Inshore Environmental Monitoring networks. After a revision of monitoring sites in 1996, more sites were added in estuaries, bays, and areas adjacent to big and medium-sized cities. Samples were collected from 1,500 monitoring sites and levels of major pollutants determined (Li, 1998). Results of pollution monitoring over the years indicated that the water quality in Chinese waters overall was basically good, but that the coastal waters were polluted to varying degrees (News Office of the State Council, 1998).

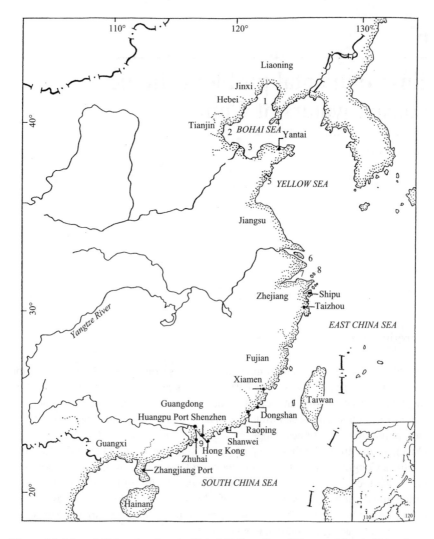

Figure 6.1 Map of China showing the Bohai, Yellow, East China, and South China Seas, and other localities
1 Liaoning Bay
2 Bohai Bay
3 Laizhou Bay
4 Dalian Bay
5 Jiaozhou Bay
6 Yangtze River estuary
7 Hangzhou Bay
8 Zhoushan fishing ground
9 Pearl River estuary
Source: Redrawn from Zhou and Wang, 1994

Polluted coastal areas have been increasing year after year. The waters of estuaries, bays, and harbours, such as those of Liaodong Bay, west Bohai Bay, Dalian Bay, Jiaozhou Bay, the Yangtze River estuary, Hangzhou Bay, the Zhoushan fishing ground, and the Pearl River estuary, were especially polluted. Data collected since 1996 present a very definite situation, one that cannot be captured with reference only to the monitoring results of earlier years.

Major anthropogenic pollutants

According to data obtained by the Coastal Environmental Monitoring Network, the major anthropogenic pollutants found in Chinese coastal waters were inorganic nitrogen, inorganic phosphorus, and oil (Wei and Zang, 1996). Concentrations of inorganic nitrogen in the Bohai, Yellow, East China, and South China Seas have exhibited a tendency to increase since 1986 (Table 6.1). In the South China Sea, inorganic nitrogen pollution has become more serious in the 1990s. Recent concentrations of inorganic nitrogen in more than 60 per cent of sites exceeded grade 1 seawater quality standards (100 μg/L). Inorganic nitrogen levels in the East China Sea were higher compared with those in the three other seas. In 1997, the mean value was more than 12 times beyond grade 1 seawater quality standards, and the percentage of monitoring sites that exceeded the standard for inorganic nitrogen levels was sometimes as high as 96 per cent.

Inorganic phosphorus contamination was most advanced in the East China Sea. In 1962 levels of inorganic phosphorus exceeded grade 1 (15 μg/L) in 62 per cent of sites, while in 1997 84 per cent of sites exceeded standards. In the Bohai and Yellow Seas, mean values were lower than the standards in 1986, 1991, and 1994, but had climbed beyond them in 1997 (Table 6.2).

Mean values of oil in Chinese seas were lower than the grade 1 seawater quality standards (0.05 mg/L) in 1986, 1991, and 1994, except those in the South China Sea in 1986. Values in the Bohai, Yellow, and East China Seas were equal to or exceeded the standards in 1997 (Table 6.3). The Bohai Sea was heavily polluted by oil in 1997, with 63 per cent of sites exceeding grade 1 seawater quality

Table 6.1 Concentrations of inorganic nitrogen in Chinese coastal waters

Year	Bohai Sea		Yellow Sea		East China Sea		South China Sea	
1986	36	11	65	29	303	78	13	19
1991	123	45	102	33	440	76	235	51
1994	137	54	138	44	470	84	344	65
1997	200	66	196	65	1,212	96	355	61

Mean value per year in μg/L, in first column; percentage of monitoring sites exceeding grade 1 seawater quality standards in second column.

Table 6.2 Concentrations of inorganic phosphorus in Chinese coastal waters

Year	Bohai Sea		Yellow Sea		East China Sea		South China Sea	
1986	7	7	5	14	14	62	6	17
1991	12	33	11	15	24	68	14	44
1994	13	35	11	55	22	66	12	33
1997	31	68	20	56	28	84	13	33

Mean value per year in μg/L, in first column; percentage of monitoring sites exceeding grade 1 seawater quality standards in second column.

Table 6.3 Concentrations of oil in Chinese coastal waters

Year	Bohai Sea		Yellow Sea		East China Sea		South China Sea	
1986	0.015	1.4	0.030	33	0.016	23	0.056	54
1991	0.014	2	0.020	5	0.032	21	0.043	27
1994	0.020	4	0.040	18	0.020	7	0.040	24
1997	0.100	63	0.050	29	0.180	20	0.040	17

Mean value per year in mg/L, in first column; percentage of monitoring sites exceeding grade 1 seawater quality standards in second column.

standards (0.05 mg/L) for oil. The percentage fluctuated between 1986 and 1997 in the other three seas.

Chinese coastal waters held low concentrations of mercury, copper, lead, and cadmium. Water quality on the whole was in accord with quality standards in terms of these heavy metals. Concentrations of some heavy metals, however, found in the sediments of some areas, are extremely high. In sea-bottom mud samples taken near the mouth of the Wulihe River of Jinzhou Bay, zinc was about 2,000 times that of the standards, while lead, cadmium, and mercury exceeded the standard by about 300, 200, and 150 times respectively.

Pollution assessment data indicated, however, that the quality of the waters off Chinese coasts showed an overall tendency to decline. There was a drop in water quality in the following sea areas: Bohai Bay; Laizhou Bay; the northern Yellow Sea; Jiaozhou Bay; off the Jiangsu coast; Hangzhou Bay; the Yangtze River estuary; the Zhoushan fishing ground; and the Pearl River estuary. Among these, the quality of the waters off the Jiangsu coast, the Yangtze River estuary, and Hangzhou Bay dropped critically and were designated as grade 3 seawater. Major pollutants exceeding standards in these areas were inorganic nitrogen, inorganic phosphorus, oil, and the like (Xinhua News Agency, 1998).

Sources of pollutants

More than 80 per cent of the pollutants in Chinese waters derive from land runoff. The coastal provinces of China are the nation's most economically developed.

Though they represent only about 13.2 per cent of the nation in terms of area, they nevertheless support approximately 40 per cent of the country's population and are responsible for more than 60 per cent of its gross national product (GNP). In the course of economic development, rapid industrialization and urbanization occurred most conspicuously along the coast. As a result, large amounts of waste material and wastewater were discharged into the seas. The volume of waste material dumped into China's seas was about 3.5 million cubic metres per year in the 1950s. This volume rose to 8 million cubic metres per year in the 1960s, 20 million cubic metres per year in the 1970s, and 40 million cubic metres per year in the 1980s. It has doubled every decade in the last 30 years. In 1994, the volume of dredged material alone was as great as 52 million cubic metres. More than 800,000 tonnes of chemical fertilizer and 200,000 tonnes of agricultural chemicals were used in coastal areas yearly, 50 per cent of which entered the seas via rivers. There are more than 200 sources of pollutants in China's coastal areas, including industrial sewage-pipe discharge, oil depots, and oil pipes, which leak or otherwise release more than 100,000 tonnes of oil into the sea every year. The total volume of industrial and municipal polluted waters entering the sea was about 6.5 billion tonnes per year in the beginning of the 1980s. This reached 8 billion tonnes in 1990, with an average increase of 0.15 billion tonnes per year (Ma, Li, and Zhang, 1996). Today, China's coastal industries and cities release up to 10 billion tonnes of wastewater containing 1.46 million tonnes of harmful toxic material into the sea each year (Zou, 1998). Most seriously polluted is the marine environment of the Bohai Sea. Here, sewage outfalls release up to 700,000 tonnes of pollutants per year, accounting for about half of the total pollutants discharged into Chinese coastal waters (Wu, J., 1998). Areas of the Bohai Sea exceeding standards expand each year. In 1995, pollutant concentrations exceeded standards in a 43,000 km^2 area, about 56 per cent of the total area of the Bohai Sea. Circumstances have yet to improve.

Pollution monitoring indicates that environmental contamination in areas adjacent to the mouths of major rivers and sewage outfalls is especially serious. Mean values of inorganic nitrogen concentrations at the mouth of the Wulihe River in Jinxi county, Liaoning province, were, in 1992 and 1993 respectively, up to 23,190 and 17,290 μg/L; those of oil were 29.53 and 45.35 mg/L. Mean levels of inorganic nitrogen in waters at the mouths of 11 rivers in Yantai city, Shandong province, were, in 1992 and 1993 respectively, 48,240 and 30,370 μg/L; those of inorganic phosphorus were 3,600 and 624 μg/L; those of oil 2.54 and 2.93 mg/L. Both Jinxi and Yantai are coastal cities along the Bohai Sea.

The volume of inorganic nitrogen and oil entering the South China Sea through the Pearl River estuary was 389,864 and 14,403 tonnes respectively in 1992, and 708,048 and 16,168 tonnes in 1993. In 1994, aerial monitoring surveys found severe oil pollution at the mouth of the Pearl River and neighbouring Huangpu Port, at Humen water course, Macao, and Zhanjiang Port, where the rate of floating oil reached 80 per cent (Wei and Zang, 1996).

An increase in tanker traffic has brought a corresponding increase in navigation accidents. About 3,000 tonnes of petroleum leaked into the sea during the *Oriental Ambassador* spill near Qingdao Port in November 1983, killing fish, invertebrates, algae, and cultivated fish in Jiaozhou Bay and up to 230 km away. Offshore petroleum production accidents have also occurred over the past 10 years. In July 1988 the Bohai No. 7 drilling platform blow-out lasted 28 hours, leaking several hundred tonnes of crude oil into Bohai Bay (Xu and Song, 1991). From the early 1970s until 1989, more than 20,000 tonnes of crude oil from over a hundred spills of more than 100 tonnes was released into the seas (Wei, Guan, and Zhan, 1995).

The drastic decline of fishery resources

Water pollution is a major cause of the decline of fishery resources in China's coastal waters. The Bohai Sea was once a major fishing ground. Now, no fishing season exists there because of pollution and overfishing. Shrimp (*Acetes* sp.) and swimming crab (*Portunus trituberculatus*) populations have declined dramatically, and no favourable return is indicated. To save the Bohai Sea is a task that brooks no delay. Likewise, during the 1970s Dalian Bay in the Yellow Sea yielded 15,000 kg of sea cucumber (*Protankura bidentata*), over 100,000 kg of scallop (*Chlamys farreri*), and more than 100,000 kg of kelp (*Laminaria japonica*) per year. Both the sea cucumber and the scallop, however, have disappeared, and kelp cultivation ceased in the 1980s due to heavy pollution. In Shandong province, the 1989 Huangdao oil depot blow-out leaked more than 630 tonnes of petroleum products into Jiaozhou Bay, contaminating 400 ha of prawn ponds, 73 ha of mussel breeding areas, and 1,167 ha of aquacultural developments, the value of the loss being estimated at over CNY 20 million (over US$240,000). One hundred and seventy species were recorded in the north-west part of Jiaozhou Bay in the 1970s, but only 17 were found in 1989 (Miao and Guan, 1996). The Yangtze estuary was a traditional icefish ground that, during the 1960s, yielded over 300 tonnes of icefish (*Salanx* sp.) per year. When industrial effluent from sewer outfalls began entering the estuary in 1971, yields declined to such an extent that the ground disappeared in the 1980s. The volume of wastewater discharged through the Yangtze estuary into the East China Sea is over 2 billion tonnes per year, a major threat to the Zhoushan fishing ground, the largest fishing ground in China (Ma, Li, and Zhang, 1996). Also in the East China Sea, a portion of Xiamen has became organism-free. The amount of industrial and domestic wastewater discharged into the South China Sea through the Pearl River estuary is about 3.7 billion tonnes per year, killing a great number of fish, shrimp, and crab. Fishery resources in the Pearl River estuary were almost eliminated due to water pollution and overexploitation.

Table 6.4 The number of red tides recorded
in coastal waters of China

Years	Number of red tides
1933–1939	1
1940–1949	0
1950–1959	1
1960–1969	3
1970–1979	15
1980–1989	208
1990–1995	Approx. 200

Red tide blooms

The earliest recorded red tide bloom in China occurred between Shipu and Taizhou, Zhejiang province, in 1933. Two hundred and twenty-eight red tides were recorded from 1933 to 1989 (Table 6.4). Among these, only five occurred before 1969, 15 occurred in the 1970s, and a sharp increase of 208 was recorded in the 1980s (Yang et al., 1994). The plague of red tides increased further in the 1990s, when nearly 200 red tides were recorded between 1990 and 1995 (Li, 1998). From autumn 1997 to spring 1998, successive attacks of red tide hit Xiamen, Dongshan, Raoping, Shanwei, Shenzhen, Zhuhai, and Hong Kong. The red tide bloom in October 1997 at Xiamen, Fujian province, was dominated by the algae species *Cyanophycium*. The major algae species found at Raoping in January 1998 was *Entophysalis aeruginosum*. Between March and April 1988, a red tide of *Gymnodinium* plagued the Pearl River estuary, including waters off Shenzhen, Zhuhai, and Hong Kong (Tang, 1998). In Hong Kong, 22 of the 26 mariculture areas were destroyed by the red tide attack, killing about 1,500 tonnes of fish, a loss of about HK$80 million. This was the worst red tide to hit Hong Kong in 10 years. In the summer of 1998, a red tide bloom at Dayugang fishing port, Wenzhou city, Zhejiang province, caused mass mortality of mullets, perches, porgies, swimming crabs, and 80 per cent of 73 ha of cultured razor clams (*Solen* sp.) (Wu, S., 1998). On 18–19 September 1998, an aerial environmental monitoring craft of the State Oceanic Administration found an extensive red tide in the Bohai Sea. Covering an area of $3,000 \text{ km}^2$ (Wu, J., 1998), it is the largest red tide recorded in Chinese waters so far. The data in Table 6.4 indicate that eutrophication levels in Chinese coastal waters are continuously becoming more serious, worsening with developments in industry and agriculture and with rapid population growth. Red tides have become more frequent and grave since the 1970s and remain a glaring environmental problem in China.

Protection of the marine environment

The Chinese government has devoted much attention to the critical situation prevailing in its coastal waters, and has made great efforts to protect the marine environment. Laws and legal provisions have been adopted and action has been taken.

In 1982, the Marine Environment Protection Law of the People's Republic of China was adopted. This is China's basic law governing protection of the marine environment, and contains legal provisions curtailing pollution from marine petroleum exploration and exploitation, ship navigation, waste dumping, and discharges from land-based sources. Regulations such as the Administrative Regulation Regarding the Prevention of Pollution from Ships, the Administrative Regulation Regarding the Environment Protection on Exploration and Exploitation of Marine Petroleum, the Administrative Regulation Regarding Ocean Waste Dumping, and the Administrative Regulation Regarding the Protection of Marine Environment from Pollutants from Land-based Sources, were promulgated.

The laws and regulations dealing with the marine environment have not been strictly enforced in some areas, however, and land-based pollution continues to increase. These two issues are the primary causes for the continued degradation of the marine environment. Thus, law enforcement and further legal provisions and measures to prevent, reduce, and control pollution of the marine environment from land-based sources have been given priority. With these efforts, we can look forward to a healthy marine environment in China in the twenty-first century.

REFERENCES

Li Bin. 1998. The circumstances of marine environment are critical. *Xinhua Daily*, 22 September 1998.

Ma Ying, Li Zongpin, and Zhang Fuchuen. 1996. Strengthening the coastal environmental monitoring for the continuous development of the coastal economy. *Marine Environmental Science*, 15(2): 57–61.

Miao Fengmin and Guan Daoming. 1996. Inquire into management of dumping area for the prevention of marine pollution. *Marine Environmental Science*, 15(2): 1–5.

News Office of the State Council. 1998. The white paper on development of ocean affairs of China. *People's Daily*, 29 May 1998, p. 5.

Tang Yongluan. 1998. The sewage discharged ocean engineering and red tides. *Shijie Keji Yanjiu Yu Fazhan*, 20(4): 117–119.

Wei Xingping and Zang Fan. 1996. Some suggestions to the marine environmental monitoring in China. *Marine Environmental Science*, 15(3): 64–70.

Wei Xingping, Guan Daoming, and Zhan Xiuwen. 1995. Some ideas to the marine environmental protection in China in the coming years. *Marine Environmental Science*, 14(4): 64–69.

Wu Jian. 1998. A large area red tide blooms in the Bohai Sea. *Workers Daily*, 4 October 1998, p. 1.

Wu Shujing. 1998. To control pollution from waters off Wenzhou is of great urgency. *China Food News*, 23 August 1998, p. 2.

Xinhua News Agency. 1998. The quality of waters off Chinese coast showed a tendency to decline. 15 May 1998.

Xu Linzhi and Song Yibao. 1991. Preliminary inquiry into counter-measures for prevention of marine pollution disasters. *Marine Environmental Science*, 10(4): 65–68.

Yang Huating, Tian Suzhen, Ye Lin, and Xu Fuxiang. 1994. *Catalog of Marine and Coastal Disasters in China (1949–1990)*, pp. 1–288. China Ocean Press, Beijing.

Yuan Li. 1998. The Bohai Sea may become a 'dead sea'. *Beijing Youth Daily*, 8 June 1998, p. 1.

Zhou Kaiya and Wang Xiaoyan. 1994. 'Brief review of passive fishing gear and incidental catches of small cetaceans in Chinese waters. *Report on the International Whaling Commission* (Special Issue), 15: 347–354.

Zou Jiahua. 1998. The water quality in Chinese coastal waters came down year after year. *Xinhua Daily*, 2 September 1998, p. 3.

7

Marine pollution monitoring of butyltins and organochlorines in the coastal waters of Thailand, the Philippines, and India

Maricar S. Prudente, Supawat Kan-Atireklap,
Shinsuke Tanabe, and Annamalai Subramanian

Introduction

The rapid increase in the buying and selling of chemicals in Asian developing countries indicates a greater production and usage of toxic chemicals, and likewise implies increased exposure of humans and wildlife to these toxins. Consequently, environmental problems caused by hazardous chemicals are of great concern in the Asia-Pacific region. The present study aims to make clear the current status of contamination by persistent toxic chemicals such as organotins and organochlorines in the coastal waters of Asian developing countries. These chemicals have been documented to have adverse effects on the environment all over the world (Tanabe, 1994; Fent, 1996).

Since the biocidial properties of trialkylated organotins were recognized in the 1950s, an increasing number of organotin compounds have been developed in industrial and agricultural commodities (Champ and Pugh, 1987). To date, butyltin compounds (BTs) representing tributyltin (TBT) were widely used for pleasure boats, large ships or vessels, docks and fish-nets, lumber preservatives, slimicides in cooling systems, and as an effective anti-fouling agent in paints. Dibutyltin (DBT) and monobutyltin (MBT) were mostly used as stabilizers in polyvinyl chloride and as catalysts in the production of polyurethane foams, silicones, and in other industrial processes (Fent, 1996). Aquatic pollution by TBT arising from anti-fouling paints has been of great concern in many countries due to its effects on non-target marine organisms, such as shell malformation in oysters (Alzieu and Heral, 1984; Alzieu and Portmann, 1984), mortality of the larvae of mussel

(Beaumont and Budd, 1984), and imposex (sterilization of females) in gastropods (Bryan and Gibbs, 1991). These effects prompted a restriction of TBT usage in many countries, for example, in the USA, France (Alzieu, 1991), the UK (Waite *et al.*, 1991; Dowson, Bubb, and Lester, 1993), Switzerland (Toth *et al.*, 1996), New Zealand (Smith, 1996), and Japan (Horiguchi *et al.*, 1994). Because of the intensive usage of BTs, they are still ubiquitous in the aquatic environment. Although TBT usage on vessels less than 25 m in length was banned, TBT-based anti-fouling paints are still being used on larger vessels, particularly commercial ships. In addition, it should be noted that use of TBT-based anti-fouling paints was mostly restricted in developed nations, whereas those in developing countries are not controlled yet. Fent (1996) suggests that BT contamination must be regarded as global pollution, particularly in countries where no regulation has been implemented, such as in Asia, Africa, and South America. In this context, an increasing demand for anti-fouling paints is predicted in the Asia-Pacific region (Layman, 1995), where aquatic pollution and toxic biological effects by BTs are anticipated. Till now, only a few studies have reported on the contamination of BTs in the Asia-Pacific region. TBT contamination has been shown in sediments, seawater, and bivalves in Hong Kong (Chiu, Ho, and Wong, 1991), Malaysia (Tong *et al.*, 1996), and Korea (Shim, 1996). BTs were detected in fish tissues and foodstuffs from several Asian and Oceanian countries (Kannan *et al.*, 1995). Imposex was also recorded in gastropods from the coastal areas of Ambon Island, Indonesia (Evans *et al.*, 1995). These observations indicate that BT contamination was widespread in Asian and Oceanian coastal waters. Despite this, comprehensive monitoring surveys of BTs have not been conducted in the Asia-Pacific coastal waters, particularly in developing countries.

Recently, organochlorine compounds (OCs) such as 1,1,1-trichloro-2,2-bis(p-chlorophenyl) ethane (DDT), polychlorinated biphenyls (PCBs), and dioxins have been postulated to show oestrogenic properties, which mimic or block the action of the natural hormone oestrogens in humans (Soto, Chung, and Sonnenschien, 1994). During the last decade, most of the developed nations have banned or restricted the production of such persistent toxic chemicals because their extensive usage has resulted in severe environmental problems and human health hazards (Tayaputh, 1996). However, OC pesticides are still being produced and widely used in developing countries for pest control in agriculture as well as for public health purposes (Siriwong *et al.*, 1991; Dave, 1996). Several studies have detected the presence of OC pesticides in various environmental compartments such as air, water, soil, sediment, and biota including green mussels from developing Asian countries (Ramesh *et al.*, 1990a, 1990b; Siriwong *et al.*, 1991; Ramesh *et al.*, 1991; Thao, 1993; Iwata *et al.*, 1994; Kanatharana, Bunvanno, and Kaewnarong, 1994; Ruangwises, Ruangwises, and Tabucanon, 1994; Kannan, Tanabe, and Tatsukawa, 1995). These studies suggest the existence of significant OC contamination in the coastal waters of tropical Asian countries in recent years. In addition, it has been pointed out that OC pesticides used in tropical regions disperse on a global scale by long-range atmospheric transport and are deposited in colder regions (Takeoka *et al.*, 1991; Iwata *et al.*, 1994). Thus, monitoring studies of OC residues in the aquatic environment of the Asia-Pacific region are needed to reduce local and global pollution problems.

Bivalves such as mussels have been addressed as bioindicators for monitoring trace toxic substances in coastal waters due to their wide geographical distribution, sessile lifestyle, easy sampling, tolerance to a considerable range of salinity, resistance to stress, and high accumulation of a wide range of chemicals (Goldberg *et al.*, 1978). Moreover, mussel has become one of the most valuable mariculture organisms produced in Asia with the largest quantity (Goldberg, 1994). The green mussel has a wide geographical distribution in the Asia-Pacific region and is a commercially valuable seafood (Vakily, 1989). Measurement of toxic contaminants in green mussels is therefore necessary in view of public health considerations.

The aim of this study is to assess the contamination by BTs and OCs in coastal waters of some Asian countries using green mussel (*P. viridis*) as a bioindicator of marine pollution. The study was conducted to determine the concentrations of TBT, DBT, and MBT in whole soft tissue of green mussels collected along coastal areas of Thailand, the Philippines, and India. OC residues such as PCBs, hexachlorocyclohexanes (HCHs), hexachlorobenzene (HCB), chlordane compounds (CHLs), and DDT including its metabolites, namely 1,1-dichlor-2,2-bis (*p*-chlorophenyl) ethane (DDD) and 1,1-dichloro-2,2 bis(*p*-chlorophenyl) ethane (DDE) were also determined.

Materials and methods

Samples

Green mussels were collected from the coastal areas of Thailand, the Philippines, and India (Figure 7.1). Samples were collected during 1994–1997. More than 30 green mussel samples were collected in each location and the adhering matrix was removed in the field. Samples were stored in polyethylene bags, kept in a cooler box with ice or dry ice, and then immediately placed in a deep freezer. The frozen green mussel samples were subsequently transported to the laboratory in Japan, thawed, scraped clean, and shucked. The whole soft tissues were pooled, homogenized, transferred to glass bottles, and then frozen at $-20°C$ until chemical analysis was carried out. The biological and site details of samples are shown in Table 7.1.

Chemical analysis

The analysis of butyltin compounds was conducted based on a procedure described in Kan-atireklap *et al.* (1997). The method consists of extraction,

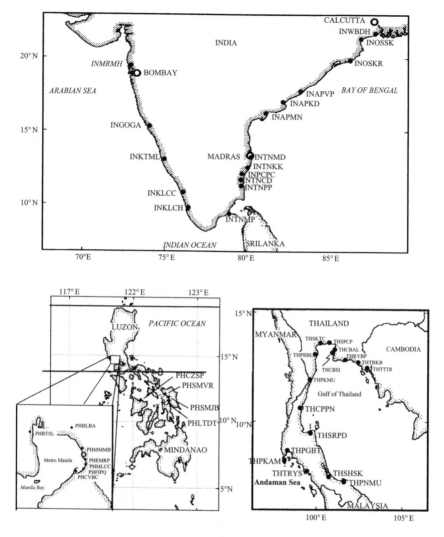

Figure 7.1 Maps showing the sampling locations of green mussels (*Perna viridis*) in India, the Philippines, and Thailand

propylation, purification, and quantification by Hewlett-Packard 5890 Series II gas chromatograph equipped with a flame photometric detector (GC-FPD). For quantification of BTs, a fused silica capillary column (J&W Scientific, 30 m length × 0.25 mm id., 0.25 mm film thickness) coated with DB-1 (100 per cent dimethyl polysiloxane) was used. Recoveries of BTs from the spiked mussel tissues were 87 ± 14 per cent, 137 ± 20 per cent, and 107 ± 25 per cent for MBT, DBT, and TBT ($n = 7$), respectively, and 3 ng g^{-1} for MBT and 1 ng g^{-1} for DBT and TBT, respectively.

Table 7.1 Locations and biological details of green mussels (*Perna viridis*) collected from the coastal waters of Thailand, the Philippines, and India

Local name	Site*	Sampling**	n	Shell length (mm)	Soft tissue weight (g)
India					
Digha, West Bengal	INWBDH	950904	21	116 (100–133)	72.4 (37.7–102.9)
Subarnarekkha, Orissa	INOSSK	950904	18	116 (90–131)	67.6 (31.4–115.0)
Konarak, Orissa	INOSKR	951116	35	58 (49–81)	13.7 (8.0–36.1)
Vishakhapatnam, Andhra Pradesh	INAPVP	950828	21	116 (101–135)	66.3 (50.0–96.2)
Kakinada, Andhra Pradesh	INAPKD	950828	29	103 (87–125)	52.8 (28.8–90.6)
Machilipatnam, Andhra Pradesh	INAPMN	950829	21	114 (91–135)	63.5 (33.0–105.8)
Madras, Tamil Nadu	INTNMD	941209	18	99 (91–111)	20.4 (15.1–26.4)
Kalpakkam, Tamil Nadu	INTNKK	950913	23	82 (61–103)	31.3 (10.2–75.3)
Pondicherry, Pondicherry	INPCPC-1	941209	14	126 (118–140)	32.1 (27.0–38.5)
Pondicherry, Pondicherry	INPCPC-2	950909	73	109 (92–136)	60.2 (31.8–102.7)
Cuddalore, Tamil Nadu	INTNCD-1	941209	28	98 (87–120)	14.9 (9.2–27.6)
Cuddalore, Tamil Nadu	INTNCD-2	950909	19	110 (94–126)	61.0 (39.1–94.1)
Parangipettai, Tamil Nadu	INTNPP	950827	23	107 (94–121)	53.8 (44.3–74.6)
Mandapam, Tamil Nadu	INTNMP	950916	48	55 (38–75)	10.7 (3.2–24.8)
Cochin, Kerala	INKLCH	951207	100	61 (50–84)	16.7 (10.5–25.1)
Calicut, Kerala	INKLCC	951206	42	107 (88–125)	90.1 (42.9–142.9)
Mangalore, Karnataka	INKTML	951205	51	94 (80–111)	58.1 (33.1–82.3)
Goa, Goa	INGOGA	951206	60	79 (38–111)	30.7 (15.4–61.4)
Mahim, Maharashtra	INMRMH	951210	43	101 (86–117)	60.7 (41.5–91.6)
Philippines					
Freedom Island, Paranaque	PHFIPQ	941122	32	40 (35–50)	1.2 (0.8–2.1)
Rizal Park, Ermita	PHEMRP	941125	19	47 (40–57)	3.0 (1.6–5.8)
CCP Complex, Malate	PHMLCC	941127	32	54 (48–63)	2.5 (1.3–4.3)
Bocaue, Bulacan	PHBLBA-1	941128	41	65 (51–80)	4.1 (1.9–8.4)
Bocaue, Bulacan	PHBLBA-2	970411	29	66 (57–80)	8.3 (4.9–13.8)
Bacoor, Cavite	PHCVBC-1	941130	39	63 (54–80)	3.6 (2.1–6.3)
Bacoor, Cavite	PHCVBC-2	970409	30	77 (63–109)	11.5 (7.1–25.6)
Jiabong, Samar	PHSMJB	970404	30	77 (48–100)	7.4 (1.6–11.3)
Villareal, Samar	PHSMVR	970405	30	76 (63–86)	8.2 (5.9–11.0)
Diit, Leyte	PHLTDT	970406	30	78 (62–114)	5.3 (1.8–14.4)
Sapian Bay, Capiz	PHCZSP	970407	18	76 (69–87)	5.1 (4.0–6.6)
Samal, Bataan	PHBTSL	970408	51	46 (35–53)	3.1 (2.0–5.4)
Malabon, Metro Manila	PHMMMB	970410	30	68 (58–80)	9.6 (5.2–15.6)
Thailand					
Trat River, Trat	THTTTR-1	940912	46	54 (49–66)	1.9 (1.3–2.7)
Trat River, Trat	THTTTR-2	950305	34	85 (34–119)	6.8 (0.9–11.4)
Kung Kra Baen, Chanthaburi	THTBKB	950306	56	49 (36–62)	1.9 (1.0–3.7)
Ban Phe, Rayong	THRYBP	940923	32	51 (37–70)	2.4 (1.0–5.6)
Sichang Island, Chonburi	THCBSI	940630	10	68 (58–76)	3.6 (2.8–4.5)

Table 7.1 (*continued*)

Local name	Site*	Sampling**	n	Shell length (mm)	Soft tissue weight (g)
Ang Sila, Chonburi	THCBAL	950307	70	43 (26–54)	1.3 (0.4–2.6)
Chao Phraya River, Samut Prakan	THSPCP-1	940904	13	69 (37–84)	3.9 (1.0–6.2)
Chao Phraya River, Samut Prakan	THSPCP-2	950323	50	65 (43–84)	5.8 (2.4–10.5)
Thachin River, Samut Sakhon	THSKTC-1	940904	18	71 (54–81)	5.6 (2.7–7.1)
Thachin River, Samut Sakhon	THSKTC-2	950322	60	77 (57–94)	6.2 (2.9–11.1)
Ban Laem, Phetchaburi	THPBBL-1	940917	16	72 (61–80)	5.5 (3.9–7.4)
Ban Laem, Phetchaburi	THPBBL -2	950312	100	37 (20–51)	0.9 (0.2–1.8)
Muang, Prachuap Khiri Khan	THPKMU-1	940915	12	79 (66–90)	4.0 (2.4–5.5)
Muang, Prachuap Khiri Khan	THPKMU-2	950312	50	58 (35–73)	3.5 (1.3–5.8)
Paknam, Chumphon	THCPPN	950321	40	81 (58–94)	6.9 (1.8–11.1)
Pak Nam Kra Dae, Surat Thani	THSRPD	950314	50	77 (59–91)	5.9 (2.7–10.6)
Songkhla Lake, Songkhla	THSHSK	950317	48	61 (47–75)	2.9 (1.1–4.6)
Muang, Pattani	THPNMU	950316	82	67 (36–84)	5.4 (1.0–9.3)
Yong Star, Trang	THTRYS	950318	30	89 (63–106)	8.4 (4.3–13.9)
Bang Toey, Phang Nga	THPGBT	950319	50	73 (54–92)	4.6 (2.2–9.9)
Ao Makham, Phuket	THPKAM	950320	70	55 (44–68)	2.5 (1.6–3.6)

*First, second and last two letters indicate the country, state, or city and local name, respectively. The digits indicate the replicate time of sampling.
**First, second and last two digits indicate the year, month, and date, respectively.
n = number of pooled samples; values in parentheses indicate the range.

PCB and OC pesticides were determined based on the analytical method of Kan-atireklap *et al.* (1998). The method consists of extraction, fat removal, acid treatment, fractionation, and quantification. Hewlett-Packard 5890 Series II gas chromatograph equipped with a ^{63}Ni electron capture detector (GC-ECD) and a moving needle-type injection system with splitless and solvent-cut mode was used for the determination of OCs. PCBs were quantified by a fused silica capillary column (30 m length $\times$ 0.25 mm id., 0.25 μm film thickness) coated with SE-54 (Supelco, USA). Quantification of OC pesticides was performed on a column consisting of fused silica capillary (30 m length $\times$ 0.25 mm id., 0.25 μm film thickness) coated with DB-1 (J&W Scientific). Total PCB concentrations in samples were quantified by summing the concentrations of individually resolved peak areas relative to an equivalent mixture standard of Aroclor (1016:1242:1254:1260; 1:1:1:1) with known PCB composition and content (Duinker, Schulz, and Petrict, 1988). Recoveries of OC residues by this method with fortified samples were 106 ± 2 per cent for PCBs, 108 ± 9 per cent for DDTs, 108 ± 6 per cent for CHLs, 105 ± 10 per cent for HCHs, and 105 ± 3 per cent for HCB ($n = 4$), respectively. The detection limit was 0.1 ng g^{-1} for PCBs and 0.01 ng g^{-1} for OC pesticides.

Results and discussion

Status of butyltin contamination

Concentrations of BTs detected in green mussels from Thailand, the Philippines, and India are summarized in Table 7.2. Among BTs, TBT was detected at all locations at relatively high concentrations, whereas the concentrations of DBT and MBT were lower. The concentrations of TBT varied widely depending on the location. Although the pollution source of BTs is unclear, these results indicate that BT contamination is widespread along the coastal waters of these countries. The concentration ranges of BTs in green mussel from Thailand were from <3 to 45 ng g^{-1} for MBT, 1 to 80 ng g^{-1} for DBT, and 3 to 680 ng g^{-1} for TBT.

Many studies have documented the existence of significant TBT contamination in harbours, marinas, shipyard hull washing/refinishing, and boating activities in developed nations (Maguire *et al.*, 1982; Page, Ozbal, and Lamphear, 1996; Tong *et al.*, 1996). TBT residues in mussels collected from boating areas (THCBSI and THTRYS) had higher TBT concentrations, suggesting that the source of TBT is from anti-fouling paints. High concentrations of TBT were also observed in green mussel collected from coastal aquaculture areas of Thailand (THTBKB, THP-KMU-1, and THPKMU-2). This implies that the usage of TBT-coated anti-fouling agents in coastal aquaculture facilities, such as fish-nets and sea pens in Thailand, is similar to that in developed countries (Short and Thrower, 1986; Davies and Mckie, 1987; Balls, 1987). Among BTs, TBT was the dominant compound, accounting for 70 per cent of the total BTs in most locations, while DBT and MBT comprised less than 20 per cent. These observations indicate that the green mussel has a limited ability to metabolize TBT to DBT and MBT, similar to other mollusks (Lee, 1986, 1991, 1995; Champ and Pugh, 1987). In addition, green mussel in the coastal waters of Thailand is exposed to a fresh input of TBT, indicating the continuous exposure to TBT in this country. Concentrations of BTs in green mussel from India ranged from <3 to 250 ng g^{-1} for MBT, <1 to 110 ng g^{-1} for DBT, and <1 to 150 ng g^{-1} for TBT. All the mussel samples collected in India contained low concentrations of BTs but detectable levels ranging from 2 to 378 ng g^{-1} as total butyltins (BTs = MBT + DBT + TBT). Relatively high concentrations of BTs were detected in samples collected from areas with intense boating (INTNMD and INOSKR), suggesting the usage of TBT as an anti-fouling agent in paints applied on boat hulls. However, concentrations of TBT in green mussel collected from coastal aquaculture areas in India (INPCPC and INTNCD) were not significant. This may indicate that the usage of TBT-coated anti-fouling agents in coastal aquaculture facilities was not

Table 7.2 Concentrations (ng g^{-1} wet wt) of butyltin compounds in green mussels (*Perna viridis*) collected from coastal waters of Thailand, the Philippines, and India

Location	MBT	DBT	TBT	BTs
India				
INWBDH	<3	1	2	3
INOSSK	16	1	1	18
INOSKR	11	12	85	108
INAPVP	20	1	1	22
INAPKD	<3	1	1	2
INAPMN	<3	1	2	3
INTNMD	9	21	150	180
INTNKK	3	1	2	6
INPCPC-1	3	2	4	9
INPCPC-2	6	1	2	9
INTNCD-1	<3	<1	3	3
INTNCD-2	24	1	1	26
INTNPP	250	110	18	378
INTNMP	11	4	24	39
INKLCH	11	3	16	30
INKLCC	8	<1	<1	8
INKTML	<3	7	10	17
INGOGA	7	1	<1	8
INMRMH	10	1	2	13
Thailand				
THTTTR-1	<3	1	8	9
THTTTR-2	<3	2	11	13
THTBKB	42	80	680	802
THRYBP	38	10	25	73
THCBSI	45	66	200	311
THCBAL	3	5	24	32
THSPCP-1	7	10	56	53
THSPCP-2	8	9	48	65
THSKTC-1	<3	3	25	28
THSKTC-2	<3	2	9	11
THPBBL-1	5	7	7	19
THPBBL-2	<3	4	23	27
THPKMU-1	9	20	130	159
THPKMU-2	7	16	210	233
THCPPN	3	3	11	17
THSRPD	4	8	49	61
THSHSK	3	5	27	35
THPNMU	5	6	41	52
THTRYS	<3	8	89	97
THPGBT	<3	1	3	4
THPKAM	<3	4	28	32

Table 7.2 (*continued*)

Location	MBT	DBT	TBT	BTs
Philippines				
PHFIPQ	15	13	76	104
PHEMRP	47	100	640	787
PHMLCC	51	43	200	294
PHBLBA-1	5	4	13	22
PHCVBC-1	9	8	34	51
PHSMJB	<3	3	1	4
PHLTDT	<3	<1	<1	nd
PHSMVR	<3	2	28	30
PHCZSP	<3	1	<1	1
PHBTSL	<3	<1	<1	nd
PHCVBC-2	<3	16	43	59
PHMMMB	3	15	44	62
PHBLBA-2	<3	12	34	46

nd = less than detection limit (3 ng g^{-1} for MBT and 1 ng g^{-1} for DBT and TBT); BTs: MBT + DBT + TBT.

so intensive in India, unlike the scenario in Thailand. BT composition in Indian mussel showed larger proportions of MBT compared to DBT and TBT in some locations. Considering the limited ability to metabolize TBT to DBT and MBT in green mussel, high MBT levels found in the present study may be due to the direct exposure to MBT and DBT from the usage of these compounds in other applications, such as heat stabilizers in polyvinyl chloride (PVC) or catalysts in industrial processes.

Concentration ranges of BTs in green mussel from the Philippines were from <3 to 51 ng g^{-1} for MBT, <1 to 100 ng g^{-1} for DBT, and <1 to 640 ng g^{-1} for TBT. The mussel samples collected in both 1994 and 1997 contained levels ranging from <3 to 787 ng g^{-1} as total BTs. Similar to Thailand and India, the green mussel from the Philippines also showed high TBT concentrations in areas with high boating activities, as in Manila Bay (PHEMRP and PHMLCC). On the other hand, levels of TBT were low in green mussel collected from aquaculture areas (PHSMJB, PHLTDT, PHCZSP, and PHBTSL). This suggests that the usage of BTs in aquaculture activities is minimal in the Philippines. When compared to the maritime or commercial vessel areas, lower TBT concentrations in green mussel were detected from several locations associated with small boats (PHEMRP and PHMLCC). Probably, the major BT pollution in the Philippines is from ocean-going sea vessels or commercial vessels, similar to that in Thailand. Among BTs, TBT was detected at most locations at relatively high concentrations, whereas the levels of DBT and MBT were lower.

Status of organochlorine contamination

Generally, concentrations of OC in green mussel were low and with small variation among locations in Thailand, the Philippines, and India (Table 7.3).

Table 7.3 Concentrations (ng g^{-1} wet wt) of OC residues in green mussels (*Perna viridis*) from the coastal waters of Thailand, the Philippines, and India

Location	Fat content (%)	PCBs	DDTs	CHLs	HCHs	HCB
India						
INWBDH	1.4	3.3	3.5	0.29	2.0	<0.01
INOSSK	1.6	2.2	1.6	0.10	3.4	<0.01
INOSKR	1.9	9.0	13.0	0.88	6.6	<0.01
INAPVP	1.7	3.0	5.0	0.66	4.0	<0.01
INAPKD	1.7	0.8	7.3	1.7	3.0	<0.01
INAPMN	1.9	1.8	6.1	0.29	3.1	<0.01
INTNMD	2.0	0.42	40.0	1.9	5.6	0.30
INTNKK	0.68	1.1	0.93	<0.01	1.5	0.22
INPCPC-1	2.5	6.1	29.0	0.15	6.6	0.12
INPCPC-2	2.3	15.0	23.0	0.32	7.6	<0.01
INTNCD-1	2.1	3.5	6.2	<0.01	4.3	0.38
INTNCD-2	1.6	0.7	12.9	0.22	3.9	<0.01
INTNPP	3.9	0.31	15.0	0.07	12.0	<0.01
INTNMP	1.3	0.38	2.1	0.19	4.5	<0.01
INKLCH	2.6	1.3	17.0	0.83	7.6	0.01
INKLCC	2.5	6.4	4.8	0.27	5.4	0.020
INKTML	2.4	3.9	18.0	0.68	10.0	0.01
INGOGA	1.6	5.1	5.8	0.17	9.5	0.02
INMRMH	2.7	2.1	7.1	0.23	10.0	0.03
Philippines						
PHFIPQ	3.1	36.0	3.3	9.5	0.19	0.04
PHEMRP	2.0	32.0	4.2	7.2	0.11	0.02
PHMLCC	1.6	33.0	3.2	7.4	0.09	0.02
PHBLBA-1	1.5	8.7	1.3	2.9	0.13	0.02
PHBLBA-2	1.8	22.0	1.7	2.5	0.15	0.01
PHCVBC-1	1.2	22.0	1.6	2.4	0.09	0.02
PHCVBC-2	1.8	14.0	1.1	3.6	0.13	<0.01
PHSMJB	1.5	0.69	1.3	0.19	<0.01	0.01
PHSMVR	1.8	2.3	0.77	0.36	0.06	0.01
PHLTDT	0.50	1.1	0.19	0.15	<0.01	0.01
PHCZSP	1.2	0.91	0.27	0.22	0.08	0.02
PHBTSL	1.7	2.6	1.6	0.34	0.16	0.01
PHMMMB	2.6	31.0	2.0	3.2	0.13	0.01
Thailand						
THTTTR-1	1.7	2.3	9.8	0.30	<0.01	<0.01
THTTTR-2	1.2	0.15	6.8	0.47	0.31	0.03

Table 7.3 (*continued*)

Location	Fat content (%)	PCBs	DDTs	CHLs	HCHs	HCB
THRYBP	1.2	9.5	38.0	5.9	<0.01	0.06
THCBSI	2.7	12.0	1.3	0.44	<0.01	0.05
THSPCP-1	2.2	3.7	5.0	2.9	0.16	0.09
THSPCP-2	1.8	20.0	7.7	3.5	0.22	0.12
THSKTC-1	1.5	0.77	2.1	0.99	0.11	0.07
THSKTC-2	1.8	0.63	1.2	0.32	0.13	0.04
THPBBL-1	1.5	0.17	2.1	0.40	0.09	0.06
THPBBL-2	1.7	2.3	6.7	2.2	0.23	0.10
THPKMU-1	2.9	0.27	2.3	0.68	0.22	0.06
THPKMU-2	2.1	0.12	2.7	0.70	<0.01	<0.01
THTBKB	1.3	1.7	3.9	0.56	0.22	0.10
THCBAL	1.8	0.40	1.3	0.37	0.10	<0.01
THCPPN	2.4	0.13	2.2	0.25	0.26	0.03
THSRPD	1.3	1.5	4.1	2.2	0.15	0.10
THSKSK	2.1	<0.1	1.4	0.30	0.26	<0.01
THPNMU	1.3	<0.1	1.9	0.93	<0.01	<0.01
THTRYS	1.2	2.0	2.4	1.3	0.33	0.09
THPGBT	1.9	0.35	14	0.28	0.14	<0.01
THPKAM	1.3	0.90	1.5	0.27	0.09	0.04

DDTs = o,p'-DDT + p,p'-DDT + p,p'-DDD + p,p'-DDE;   CHLs, t-chlordane + c-chlordane + t-nonachlor + c-nonachlor; HCHs, α-HCH + β-HCH + γ-HCH.

Concentrations of PCBs in green mussel from the coastal waters of Thailand exhibited a smaller variation among locations, ranging from 0.17 to 12 ng g^{-1} in 1994 and <0.1 to 20 ng g^{-1} in 1995. Relatively high levels of PCBs were detected at some locations around the most populated and industrialized cities (THSPCP and THCBSI). One of the major PCB sources in Thailand is considered to have originated from transformers and capacitors, which were imported by the Electricity Authority of Thailand. Watanabe *et al.* (1996) documented PCB pollution in the dumping site of transformers and capacitors located at the suburbs of Bangkok (near the estuary of the Chao Phraya River). This fact raises concern about the possible increase of PCB pollution in Thailand.

Besides PCBs, DDT residues were detected with the highest concentration in green mussel from Thailand, followed by CHLs, HCHs, and HCB. DDT and its metabolites have been found in mussels along coastal waters of Thailand, ranging from 1.3 to 38 ng g^{-1} and 1.2 to 15 ng g^{-1} in 1994 and 1995, respectively. The levels of DDTs in green mussel were similar to those in fish collected from Thailand, ranging from 0.48 to 19 ng g^{-1} (Kannan, Tanabe, and Tatsukawa, 1995). In Thailand, the usage of DDT for agriculture purposes has been banned since 1983, but is still being used for malaria vector control (Ruangwises, Ruangwises, and Tabucanon, 1994). Considerable residues of p,p'- DDT in DDTs found in green mussel from some locations may indicate the presence of significant current sources of DDT in Thailand. However, based on the results shown in earlier studies (Menasveta and Cheevaparanapiwat, 1981; Siriwong et al., 1991; Ruangwises, Ruangwises, and Tabucanon, 1994), it appears that the concentrations of DDT have declined in Thailand coastal waters since the restriction was introduced in 1983.

Among the OC pesticides examined, CHLs were detected at levels next to DDT, ranging from 0.3 to 5.9 ng g^{-1} in 1994 and 0.25 to 3.5 ng g^{-1} in 1995. In biological samples other than mussels, CHL data are available for fish (Kannan, Tanabe, and Tatsukawa, 1995), which showed a similar range of concentration from 0.1 to 15 ng g^{-1} in Thailand. Concentrations of HCHs were relatively low, ranging from <0.01 to 0.33 ng g^{-1}. α-HCH was detected as the predominant isomer in most locations. Kannan, Tanabe, and Tatsukawa (1995) also found α-HCH as a predominant residue in fish from Thailand. These results may reflect the past usage of technical HCH in Thailand. The usage of technical HCH was banned in the 1980s, nevertheless the application of γ-HCH (lindane) still appears to be continuing in Thailand (Siriwong et al., 1991). The detection of γ-HCH as a prevalent isomer in some locations (THSKSK and THTRYS) suggests the usage of lindane in Thailand. HCB contamination seems to have originated from the usage of fungicides or as an impurity in pesticide formulations, a byproduct of various chlorination processes and the combustion of industrial and municipal wastes (Kannan, Tanabe, and Tatsukawa, 1995).

The concentrations of HCB found in the present study ranged from <0.01 to 0.12 ng g^{-1}, which was similar to that found in earlier studies in 1989, ranging from ND to 0.12 ng g^{-1} (Siriwong et al., 1991), and in 1991, ranging from 0.01 to 0.21 ng g^{-1} (Ruangwises, Ruangwises, and Tabucanon, 1994). The spatial differences in OC residue levels in green mussels collected from several locations along the coastal waters suggest that these compounds were uniformly distributed in Thailand. Although the agricultural usage of OC pesticides, such as DDT and HCH, was banned in Thailand, some other sources still remain in the industrial and human activities. When compared with previous data (Menasveta and Cheevaparanapiwat, 1981; Siriwong et al., 1991; Ruangwises, Ruangwises, and Tabucanon, 1994), residue levels of OC pesticides in Thailand's marine environment have been declining. Considering the low levels of OCs in green mussels, however, contamination is not so serious from human health and ecosystem perspectives.

Concentrations of PCBs in green mussels collected from the Indian coast were low and exhibited a smaller variation among locations, ranging from 0.31 to 15 ng g^{-1}. Low PCB levels have been detected in human breast milk (Tanabe et al., 1990) and green mussel from the coastal waters of South India (Ramesh et al., 1990b). Likewise, low levels of PCBs were found in fish muscle, averaging 3.5 ng g^{-1} wet weight, from India (Kannan, Tanabe, and Tatsukawa, 1995). Comparison of the concentrations of PCBs in Indian green mussels with those of earlier reports indicated that the inputs of PCBs are still occurring in Indian coastal waters. In general, DDT concentrations in Indian mussels were the highest among various OC pesticides analysed in this study, followed by HCHs, CHLs, and HCB, respectively. The range of DDT concentrations observed in the present study was from 6.2 to 40 ng g^{-1} and 0.93 to 22 ng g^{-1} in 1994 and 1995, respectively. The concentrations of DDTs in green mussel were comparable to or lower than those in fish collected from India, which contained 0.86 to 140 ng g^{-1} (Kannan, Tanabe, and Tatsukawa, 1995). Compared with earlier studies (Ramesh *et al.*, 1990b), DDT contamination in Indian coastal waters is likely to be constant.

These results suggest that DDT is still being produced and used in India. High levels of DDTs were found in some urban locations (INOSKR, INTNMD, and INPCPC), which may be an indication of contamination originating from its use for public health purposes. During 1994 and 1995, 4,252 tonnes of DDT were produced in India (Dave, 1996). Moreover, 85 per cent of the total annual consumption of DDT in India was used for mosquito control (Singh, Battu, and Kalra, 1988). A considerably higher ratio of *p,p'*- DDT in total DDTs (mean = 33 per cent) may support the current usage of DDT in India for public health purposes.

The concentrations of HCHs in green mussels from India were in the range of 1.5–12 ng g^{-1}, which were comparable to that of an earlier study in 1988–1989 (Ramesh *et al.*, 1990b). Among HCH isomers, α-HCH was detected as the predominant isomer in most locations (mean = 67 per cent). This result may reflect the recent usage of a technical mixture of HCH in India. Kannan, Tanabe, and Tatsukawa (1995), based on an analysis of fish and foodstuffs, also reported the predominance of α-HCH in India. Recent reports indicated that India has been a major source of HCH production in the world, amounting to 15,099 tonnes in 1980, 19,880 tonnes in 1990 (Li, McMillan, and Scholtz, 1996), and 32,026 tonnes during 1994–1995 (Dave, 1996). On the other hand, low levels of γ-HCH were found in green mussels, suggesting a smaller usage of lindane. This was further evidenced by a report on the usage of 4,000 tonnes a year from 1988 to 1992 (Li, McMillan, and Scholtz, 1996).

Concentrations of CHLs in Indian green mussel were relatively low, ranging from <0.01 to 1.9 ng g^{-1} in 1994 and <0.01 to 1.8 ng g^{-1} in 1995. Although the data on CHL residues in bivalves are scarce, the residue levels were comparable to earlier studies (Ramesh *et al.*, 1990b). Concentrations of HCB were the lowest, ranging from <0.01 to 0.38 ng g^{-1}, among OCs examined. This suggests the limited use of HCB in India. The residue levels were comparable to those in Indian fish, ranging from <0.01 to 0.55 ng g^{-1} (Kannan, Tanabe, and Tatsukawa, 1995). Generally, spatial variation in OC levels in green mussels collected along the coastal waters of India suggested that these compounds have been widely distributed.

Concentrations of PCBs in green mussels from the coastal waters of the Philippines were detected in a range of 8.7 to 36 ng g^{-1} in 1994 and 0.69 to 31 ng g^{-1} in 1997. The PCB contamination pattern was similar to that in Thailand mussels. These results imply that PCB contamination comes from waste drainage from the most populated and industrialized cities. Unlike mussels from India and Thailand, CHL concentrations in mussels from the Philippines were the highest among various OC pesticides determined in the present study, followed by DDTs, HCH, and HCB. Concentrations of CHLs in green mussels were detected in the range of 2.4 to 9.5 ng g^{-1} in 1994 and 0.15 to 3.6 ng g^{-1} in 1997. The concentration of HCHs were relatively low, ranging from 0.09 to 0.19 ng g^{-1} in 1994 and <0.01 to 0.16 ng g^{-1} in 1997. HCB concentrations were the lowest, ranging from 0.02 to 0.04 ng g^{-1} and <0.01 to 0.02 ng g^{-1} in 1997.

Geographical distribution of butyltin and organochlorine contamination

As a whole, the highest level of BTs was found in mussels from the Philippines, followed by Thailand and India. BT levels in mussels collected from aquaculture areas in Thailand implied that TBT had been used for some equipment in aquaculture activities. These observations suggest the possibility that the usage of TBT for aquaculture activities may increase in developing countries with a high economic growth rate, for example Thailand. High BT contamination in mussels was observed in areas of high boating activities, particularly maritime vessels or large commercial vessels and fishing boat areas in Thailand and the Philippines, suggesting the current usage of TBT-coated anti-fouling agents on ship hulls.

As seen in Figure 7.2, relatively high levels of PCBs were found in green mussels collected from the Philippines, while lower levels were found in Thailand and

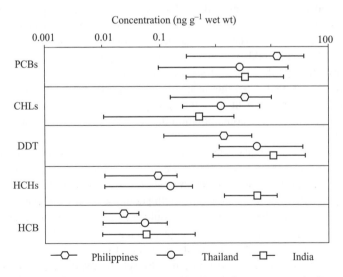

Figure 7.2 Comparison of organochlorine residue levels in green mussels (*Perna viridis*) from India, the Philippines, and Thailand

India. However, sporadic incidences of higher levels of PCBs were observed at urbanized and industrialized areas in all the countries surveyed. Among OC pesticides determined, the concentration levels of DDTs and HCHs were relatively higher in Indian mussels (Figure 7.2). However, high levels of DDTs and HCHs were occasionally detected in mussels from some locations of Thailand and the Philippines, which might be indicative of the continuing usage of these insecticides for public health and aquacultural purposes. Collectively, OC residues in green mussels from Asian coastal waters suggest that these compounds are widely used in the Asia-Pacific region.

International comparison of butyltin and organochlorine contamination

The levels of BTs in green mussels collected from India, Thailand, and the Philippines in the present study were compared to those in developed nations (Table 7.4). Among the Asian developing countries examined, BT levels in green mussels from the coastal waters of the Philippines and Thailand were higher than those in India, Malaysia, and Hong Kong. Although the monitoring data are scarce in the Asian regions, pollution by BTs in the Philippines and Thailand may be ranked as "high". When compared with recent surveys in other bivalve species from various locations, the BT levels in green mussels from India, Thailand, and the Philippines were lower than those in developed nations including Japan and European and North American countries.

Although the BT contamination levels in Asian countries were lower than those in developed nations, the demand for TBT-based marine coating paints in Asia and Oceania is presently increasing. The average annual growth rate of the demand for anti-fouling paints in Asia and Oceania was predicted to be 5.7 per cent from 1995 to 2000, which is the highest in the world (Reisch, 1996). Considering the fact that there is unregulated usage of BTs in India, Thailand, and the Philippines as well as in other nearby developing countries, BT contamination in Asian aquatic environments may increase in the future. Continuous monitoring and investigations on the contamination and effect of BTs are needed in Asian developing nations.

The international comparison of OC residues in green mussels from India, Thailand, and the Philippines with those in other countries is summarized in Table 7.5. The levels of OC residues in green mussels found in the present study were comparable to those of other developing countries, but were relatively lower than those of developed nations. This could be explained by a previous flux modelling study in the tropics, which suggested that the OC budget to the aquatic ecosystem was less significant and its residence time in the environment was quite short, whereas the OC transfer to the atmosphere was much larger due to the high temperature in the tropics (Takeoka *et al.*, 1991). These features were confirmed in earlier published reports on OC fate and behaviour from a global

Table 7.4 Concentrations of butyltin compounds in bivalve molluscs collected worldwide

Species	Location	Concentration (ng g^{-1} wet wt)			References
		MBT	DBT	TBT	
Mytilus edulis	Pacific coast, USA, 1986–1987	NA	NA	<5–1,080	Short and Sharp, 1989
	Tokyo Bay, Japan, 1989	20–120	40–540	20–240	Higashiyama et al., 1991
	East coast, USA	ND–140	10–580	10–1,200	Uhler et al., 1993
	West coast, USA	ND–300	10–740	10–1,380	Uhler et al., 1993
	British Columbia, Canada, 1990	8.2–49	6.8–80	52–314	Stewart and Thompson, 1994[a,c]
	Maine, USA, 1989	NA	340 ± 50	2,350 ± 430	Page, Dassanayake, and Gilfillan, 1995[a]
	Perth, Australia, 1991	NA	NA	<1–330	Burt and Ebell, 1995
Mytilus galloprovincialis	West coast, Portugal	3.2–169	ND–82	ND–114	Quevauviller et al., 1989
	Mediterranean, 1988	NA	12–2,450	220–2,600	Tolosa et al., 1992[a]
Mya areanaria	Fyn, Denmark, 1989	NA	NA	250–1,470	Kure and Depledge, 1994
	Poole Harbour, UK, 1987	NA	1,780–9,950	5,550–11,450	Langston, Burt, and Mingjiang, 1987[a,b]
Thais clavigera	Japan	0–1,100	0–1,200	0–380	Horiguchi et al., 1994[d]
Thais bronni	Japan	0–560	0–1,120	0–720	Horiguchi et al., 1994[d]
Crassostrea virginica	Gulf of Mexico, 1989	<5–145	<5–380	<5–1,450	Garcia-Romero et al.,1993[a,b]
	Gulf of Mexico, 1990	<5–25	<5–160	<5–770	Garcia-Romero et al., 1993[a,b]
	Gulf of Mexico, 1991	<5–42	<5–200	<5–1,160	Garcia-Romero et al., 1993[a,b]
	East coast, USA	ND–330	10–1570	10–4,030	Uhler et al., 1993
	Chesapeake Bay, USA	NA	ND	<10–5,600	Espourteille, Greaves, and Huggett, 1993
Saccostrea commercialis	Hawkesbury River estuary, Australia, 1991	NA	NA	ND–1,700	Hardiman and Pearson, 1995

Table 7.4 (*continued*)

Species	Location	Concentration (ng g⁻¹ wet wt)			References
		MBT	DBT	TBT	
Crassostrea gigas	Coastal estuaries, UK, 1986	NA	NA	180–6,350	Waite *et al.*, 1991
	Coastal estuaries, UK, 1987	NA	NA	280–3,650	Waite *et al.*, 1991
	Coastal estuaries, UK, 1988	NA	NA	80–5,600	Waite *et al.*, 1991
	Coastal estuaries, UK, 1989	NA	NA	80–1,280	Waite *et al.*, 1991
	Chinhae Bay, Korea, 1995	19–119	53–559	250–1,350	Shim, 1996[a,b]
Perna viridis	Hong Kong, 1989	NA	NA	64–115[c]	Chiu, Ho, and Wong, 1991
	Malaysia, 1992	NA	NA	14.2–23.5[c]	Tong *et al.*, 1996
	India, 1994–1995	<3–250	<1–110	<1–150	This study
	Philippines, 1994–1997	<3–51	<1–100	<1–640	This study
	Thailand, 1994–1995	<3–45	1–80	3–680	This study

[a] Dry weight basis.
[b] Express in Sn.
[c] Mean.
[d] Determined from graph figures.
NA: not analysed; ND: no data shown.

Table 7.5 Concentrations (ng/g^{-1} wet wt) of organochlorine residues in bivalve molluscs collected worldwide

Species	Location	PCBs	DDTs	CHLs	HCHs	HCB	Reference
Mythus edulis	Kattegat, Denmark, 1985	3–328	24–67	—	0.6–7.4	—	Granby and Spliid (1995)
	Perth, Western Australia, 1991	<10	<1–2	—	—	—	Burt and Ebell (1995)
	North-west Spain, 1990–1991	nd–620	nd–36	—	nd–49	—	Alvarez Piñeiro, Simal Lozano, and Lage Yusty (1995)
	North, Central, and South America, 1986–1993	10–380	6.7–960	2.8–9.6	—	—	Sericano et al. (1995)[a]
	Coastal water of USA, 1992	0	4	1.2	—	—	Lauenstein (1995)[b]
	Coastal water of USA, 1993	12	5	1.4	—	—	O'Connor (1996)[b]
		16	1–18	—	0.3–4.5	nd–1	Lee, Kruse, and Wassermann (1996)
	South-west Baltic Sea, 1990–1991	4.7–97					
Perna viridis	Hong Kong, 1986	49–330	50–520	—	53–110	—	Phillips (1989)
	South India, 1988–1989	0.66–7.1	2.8–40	—	4.3–16	—	Ramesh et al. (1990b)
	Coastal, India, 1994–1995	0.31–15	0.93–40	<0.01–1.9	1.5–12	<0.01–0.38	This study
	Coastal water of Thailand, 1989	—	0.39–7.41	—	<0.02–0.19	<0.02–0.31	Siriwong et al. (1991)
	Coastal water of Thailand, 1991	—	0.74–5.38	—	<0.02–0.09	<0.02–0.21	Ruangwises, Ruangwises, and Tabucanon (1994)
	Coastal water of Thailand, 1994–1995	<0.01–20	1.3–38	0.25–5.9	<0.01–0.43	<0.01–0.12	This study
	Philippines, 1994–1997	0.69–36	0.19–4.2	0.15–9.5	<0.01–0.19	<0.01–0.19	This study

[a] Dry weight basis of bivalves (i.e. oysters, mussels, and other bivalves).
[b] Geometric mean of *Mytilus edulis*, *Mytilus californianus*, *Crasostea virginica*.

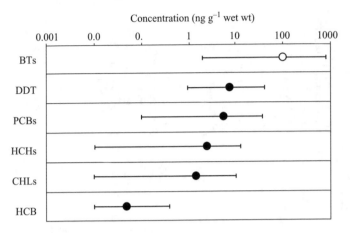

Figure 7.3 Range and mean concentrations of persistent toxic contaminants in green mussels from tropical coastal waters in Asia

perspective (Iwata *et al.*, 1994). In addition, OCs are still being produced and used in some Asian developing countries such as India, which is rapidly becoming one of the most dynamic generic producers of agrochemicals in the world because it has a low-cost manufacturing capacity (Dave, 1996). From this perspective, the continuous monitoring of OC residues in Asia is deemed necessary.

Among the persistent contaminants examined in green mussels collected from Asian developing countries, BTs showed higher residue levels than OC residues (Figure 7.3), predicting a sign of accelerating pollution and toxic effects by organotins in Asian coastal waters in the future.

Acknowledgements

The authors wish to thank Dr K. Kanan (Michigan State University, USA) for critical reading of this manuscript. Thanks are also due to Dr K. Kitazawa (UNESCO-IOC, President: Japan Marine Science and Technology Center) and Dr G. Kullenberg (UNESCO-IOC), G. Paoletto, and Dr J. I. Uitto (United Nations University, Japan), Professor E. D. Goldberg (Chairman of the International Mussel Watch Committee), Professor J. W. Farrington, and B. W. Tripp (Woods Hole Institute of Oceanography, USA) for their encouragement and support. They also would like to acknowledge the staff in the Eastern Marine Fisheries Development Center, Department of Fisheries, Environmental Research and Training Center, Department of Environmental Quality Promotion, Thailand, Centre of Advanced Study in Marine Biology, Annamalai University, India, and Science Education students of De La Salle University, Manila, Philippines for their help with sample collection. This research was supported by a grant-in-aid from the International Scientific Research Programme of the Ministry of Education, Science, and Culture of Japan (Project No. 09041163). This project was conducted on the duration of a JSPS Ronpaku Fellowship Grant (DOST 9639) awarded to M. S. Prudente.

REFERENCES

Alvarez Piñeiro, M.E., J. Simal Lozano, and M.A. Lage Yusty. 1995. Organochlorine compounds in mussels of the estuarine bays of Galicia (north-west Spain). *Marine Pollution Bulletin*, 30: 484–487.

Alzieu, C. 1991. Environmental problems caused by TBT in France: Assessment, regulations, prospects. *Marine Environment Research*, No. 32, pp. 7–17.

Alzieu, C. and M. Heral. 1984. Ecotoxicological effects of organotin compounds on oyster culture. In: *Ecotoxicological Testing for the Marine Environment*, Vol. 2, G. Persoone, E. Jaspers, and C. Claus (eds), pp. 187–195. State University, Belgium.

Alzieu, C. and J.E. Portmann. 1984. The effect of tributyltin on the culture of *C. gigas* and other species. In: *Fifteenth Annual Shellfish Conference Proceedings*, pp. 87–104. Shellfish Association of Great Britain, London.

Balls, P.W. 1987. Tributyltin (TBT) in the waters of a Scottish sea loch arising from the use of antifoulant treated netting by salmon farms. *Aquaculture*, No. 65, pp. 227–237.

Beaumont, A.R. and M.D. Budd. 1984. High mortality of the larvae of the common mussel at low concentrations of tributyltin. *Marine Pollution Bulletin*, 15: 402–405.

Bryan G.W. and P.E. Gibbs. 1991. Impact of low concentrations of tributyltin (TBT) on marine organisms: A review. In: *Metal Ecotoxicology Concepts and Applications*, M.C. Newman and A.W. McIntosh (eds), pp. 323–361. Lewis Publishers, USA.

Burt, J.S. and G.F. Ebell. 1995. Organic pollutants in mussels and sediments of the coastal waters off Perth, western Australia. *Marine Pollution Bulletin*, 30: 723–732.

Champ, M.A. and W.L. Pugh. 1987. Tributyltin antifouling paints: Introduction and overview. In: *IEEE Oceans '87 Conference Proceedings*, pp. 1296–1308. Marine Technology Society, Washington, DC.

Chiu, S.T., L.M. Ho, and P.S. Wong. 1991. TBT contamination in Hong Kong waters. *Marine Pollution Bulletin*, 22: 220.

Dave, P.P. 1996. India: A generics giant. *Farm Chemicals International*, November, pp. 36–37.

Davies, I.M. and J.C. Mckie. 1987. Accumulation of total tin and tributyltin in muscle tissue of farmed Atlantic salmon. *Marine Pollution Bulletin*, 18: 405–407.

Dowson, P.H., J.M. Bubb, and J.N. Lester. 1993. Temporal distribution of organotins in the aquatic environment: Five years after the 1987 UK retail ban on TBT-based antifouling paints. *Marine Pollution Bulletin*, 26: 487–494.

Duinker, J.C., D.E. Schulz, and G. Petrict. 1988. Multidimensional gas chromatography with electron capture detection for the determination of toxic congeners in polychlorinated biphenyl mixtures. *Journal of Analytical Chemistry*, 60: 478–482.

Espourteille, F.A., J. Greaves, and R.J. Huggett. 1993. Measurement of tributyltin contamination of sediments and *Crassostrea virginica* in the southern Chesapeake Bay. *Environmental and Toxicology Chemistry*, 12: 305–314.

Evans, S.M., J. Dawson, J. Day, C.L.J. Frid, M.E. Gill, L.A. Pattisina, and J. Porter. 1995. Domestic waste and TBT pollution in coastal areas of Ambon Island (eastern Indonesia). *Marine Pollution Bulletin*, 30: 109–115.

Fent, K. 1996. Ecotoxicology of organotin compounds. *Critical Reviews in Toxicology*, 26: 1–117.

Garcia-Romero, B., T.L. Wade, G.G. Salata, and J.M. Brooks. 1993. Butyltin concentrations in oysters from the Gulf of Mexico from 1989 to 1991. *Environmental Pollution*, No. 81, pp. 103–111.

Goldberg, E.D. 1994. *Coastal Zone Space: Prelude to Conflict?* UNESCO, 138pp.

Goldberg, E.D., V.T. Bowen, J.W. Farrington, G.R. Harvey, J.H. Martin, P.L. Parker, R.W. Risebrough, W.E. Robertson, E. Schneider, and E. Gamble. 1978. The Mussel Watch. *Environmental Conservation*, 5: 101–125.

Granby, K. and N.H. Spliid. 1995. Hydrocarbons and organochlorines in common mussels from the Kattegat and the belts and their relation to condition indices. *Marine Pollution Bulletin*, 30: 74–82.

Hardiman, S. and B. Pearson. 1995. Heavy metals, TBT and DDT in the Sydney rock oyster (*Saccostrea commercialis*) sampled from the Hawkesbury River estuary, NSW, Australia. *Marine Pollution Bulletin*, 30: 563–567.

Higashiyama, T., H. Shiraishi, A. Otsuki, and S. Hashimoto. 1991. Concentrations of organotin compounds in blue mussels from the wharves of Tokyo Bay. *Marine Pollution Bulletin*, 22: 585–587.

Horiguchi, T., H. Shiraishi, M. Shimizu, and M. Morita. 1994. Imposex and organotin compounds in *Thais clavigera* and *T. bronni* in Japan. *Journal of the Marine Biological Association*, 74: 651–669.

Iwata, H., S. Tanabe, N. Sakai, A. Nishimura, and R.Tatsukawa. 1994. Geographical distribution of persistent organochlorines in air, water and sediments from Asia and Oceania, and their implications for global redistribution from lower latitudes. *Environmental Pollution*, No. 85, pp. 15–33.

Kanatharana, P., S. Bunvanno, and B. Kaewnarong. 1994. Organochlorine pesticide residues in Songkia Lake. *Environmental Monitoring and Assessment*, No. 33, pp. 43–52.

Kan-atireklap, S., S. Tanabe, J. Sanguansin, M.S. Tabucanon, and M. Hungspreugs. 1997. Contamination by butyltin compounds and organochlorine residues in green mussel (*Perma viridis*, L.) from Thailand coastal waters. *Environmental Pollution*, No. 97, pp. 79–89.

Kan-atireklap, S., N.T.H. Yen, S. Tanabe, and A.M. Subramanian. 1998. contamination by butyltin compounds and organochlorine residues in green mussel (*Perma viridis*, L.) from India. *Environmental Toxicology and Chemistry*, No. 68, pp. 409–424.

Kannan, K., S. Tanabe, and R. Tatsukawa. 1995. Geographical distribution and accumulation features of organochlorine residues in fish in tropical Asia and Oceania. *Environmental Science and Technology*, 29: 2673–2683.

Kannan, K., S. Tanabe, H. Iwata, and R. Tatsukawa. 1995. Butyltins in muscle and liver of fish collected from certain Asian and Oceanian countries. *Environmental Pollution*, No. 90, pp. 279–290.

Kure, L.K. and M.H. Depledge. 1994. Accumulation of organotin in *Littorina littorea* and *Mya arenaria* from Danish coastal waters. *Environmental Pollution*, No. 84, pp. 149–157.

Langston, W.J., G.R. Burt, and Z. Mingjiang. 1987. Tin and organotin in water, sediments, and benthic organisms of Poole Harbour. *Marine Pollution Bulletin*, 18: 634–639.

Lauenstein, G.G. 1995. Comparison of organic contaminants found in mussels and oysters from a current Mussel Watch project with those from archived mollusc samples of the 1970s. *Marine Pollution Bulletin*, 30: 826–633.

Layman, P.L. 1995. Marine coatings industry adopts new technologies for shifting markets. *Chemical and Engineering News*, 1 May, pp. 23–25.

Lee, K.M., H. Kruse, and O. Wassermann. 1996. Seasonal fluctuation of organochlorines in Mytilus edulis L. from the south west Baltic Sea. *Chemosphere*, 32: 1883–1895.

Lee, R.F. 1986. Metabolism of bis(tributyltin) oxide by estuarine animals. In: *IEEE Oceans '86 Conference Proceedings*, pp. 1182–1188. Marine Technology Society, Washington, DC.

Lee, R.F. 1991. Metabolism of tributyltin by marine animals and possible linkages to effects. *Marine Environmental Research*, No. 32, pp. 29–35.

Lee, R.F. 1995. Metabolism of tributyltin by aquatic organisms. In: *Organotin*, M. Champ and P.F. Seligman (eds), pp. 369–382. Chapman and Hall, London.

Li, Y.-F., A. McMillan, and M.T. Scholtz. 1996. Global HCH usage with $1° \times 1°$ longitude/latitude resolution. *Environmental Science and Technology*, 30: 3525–3533.

Maguire, R.J., Y.K. Chau, G.A. Bengert, E.J. Hale, P.T.S. Wong, and O. Kramar. 1982. Occurrence of organotin compounds in Ontario lakes and rivers. *Environmental Science and Technology*, 16: 698–702.

Menasveta, P. and V. Cheevaparanapiwat. 1981. Heavy metals, organochlorine pesticides and PCBs in green mussels, mullets and sediments of rivers in Thailand. *Marine Pollution Bulletin*, 12: 19–25.

O'Connor, T.P. 1996. Trends in chemical concentrations in mussels and oysters collected along the US coast from 1986 to 1993. *Marine Environmental Research*, No. 41, pp. 183–200.

Page, D.S., T.M. Dassanayake, and E.S. Gilfillan. 1995. Tissue distribution and depuration of tributyltin for field-exposed *Mytilus edulis*. *Marine Environmental Research*, No. 40, pp. 409–421.

Page, D.S., C.C. Ozbal, and M.E. Lanphear. 1996. Concentration of butyltin species in sediments associated with shipyard activity. *Environmental Pollution*, No. 91, pp. 237–243.

Phillips, D.J.H. 1989. Trace metals and organochlorines in the coastal waters of Hong Kong. *Marine Pollution Bulletin*, 20: 319–27.

Quevauviller, P., R. Lavigne, R. Pinel, and M. Astruc. 1989. Organotins in sediments and mussels from the Sado estuarine system (Portugal). *Environmental Pollution*, No. 57, pp. 149–166.

Ramesh, A., S. Tanabe, H. Iwata, R. Tatsukawa, A.N. Subramanian, D. Mohan, and V.K. Venugopalan. 1990a. Seasonal variations of persistent organochlorine insecticide residues in Vellar River waters in Tamil Nadu, south India. *Environmental Pollution*, No. 67, pp. 289–304.

Ramesh, A., S. Tanabe, A.N. Subramanian, D. Mohan, V.K. Venugopalan, and R. Tatsukawa. 1990b. Persistent organochlorine residues in green mussels from coastal waters of south India. *Marine Pollution Bulletin*, 21: 587–590.

Ramesh, A., S. Tanabe, H. Murase, A.M. Subramanian, and R. Tatsukawa. 1991. Distribution and behaviour of persistent organochlorine insecticides in paddy soil and sediments in the tropical environment: A case study in south India. *Environmental Pollution*, No. 74, pp. 293–307.

Reisch, M.S. 1996. Paints and coatings. *Chemical Engineering News*. 14 October, pp. 44–49.

Ruangwises, S., N. Ruangwises, and M.S. Tabucanon. 1994. Persistent organochlorine pesticide residues in green mussels (*Perma viridis*) from the Gulf of Thailand. *Marine Pollution Bulletin*, 28: 351–355.

Sericano, J.L., T.L. Wade, T.J. Jackson, J.M. Brooks, B.W. Tripp, J.W. Farrington, L.D. Mee, J.W. Readmann, J.P. Villeneuve, and E.D. Goldberg. 1995. Trace organic contamination in the Americas: An overview of the US national status and trends and the international 'Mussel Watch' programmes. *Marine Pollution Bulletin*, 31: 214–225.

Shim, W.J. 1996. Contamination and bioaccumulation of tributyltin and triphenyltin compounds in the Chinhae Bay system, Korea. Msc thesis, Oceanography Department, Seoul National University, 90pp.

Short, J.W. and F.P. Thrower. 1986. Accumulation of butyltins in muscle tissue of chinook salmon reared in sea pens treated with tri-*n*-butyltin. In: *IEEE Oceans '86 Conference Proceedings*, pp. 1177–1181. Marine Technology Society, Washington, DC.

Short, J.W. and J.L. Sharp. 1989. Tributyltin in bay mussels (*Mytilus edulis*) of the Pacific coast of the United States. *Environmental Science and Technology*, 23: 740–743.

Singh, P.P., R.S. Battu, and R.L. Kalra. 1988. Insecticide residues in wheat grains and straw arising from their storage in premises treated with BHC and DDT under malaria control programs. *Bulletin of Environmental Contamination and Toxicology*, No. 40, pp. 696–702.

Siriwong, C., H. Hironaka, S. Onodera, and M.S. Tabucanon. 1991. Organochlorine pesticides residues in green mussel (*Perna viridis*) from the Gulf of Thailand. *Marine Pollution Bulletin*, 22: 510–516.

Smith, P.J. 1996. Selective decline in imposex levels in the dogwhelk *Lepsiella scobina* following a ban on the use of TBT antifoulants in New Zealand. *Marine Pollution Bulletin*, 32: 362–365.

Soto, A.M., K.L. Chung, and C. Sonnenschein. 1994. The pesticides endosulfan, toxaphene, and dieldrin have estrogenic effects on human estrogen-sensitive cells. *Environmental Health Perspectives*, No. 102, pp. 380–383.

Stewart, C. and J.A.J. Thompson. 1994. Extensive butyltin contamination in south-western coastal British Columbia, Canada. *Marine Pollution Bulletin*, 28: 601–606.

Takeoka, H., A. Ramesh, H. Iwata, S. Tanabe, A.N. Subramanian, D. Mohan, A. Magendran, and R. Tatsukawa. 1991. Fate of the insecticide HCH in the tropical coastal area of south India. *Marine Pollution Bulletin*, 22: 290–297.

Tanabe, S. 1994. International mussel watch in Asia-Pacific phase. *Marine Pollution Bulletin*, 28: 518.

Tanabe, S., F. Gondaira, A.N. Subramanian, A. Ramesh, D. Mohan, P. Kumarun, V.K. Venugopalan, and R. Tatsukawa. 1990. Specific pattern of persistent organochlorine residues in human breast milk from south India. *Journal of Agricultural and Food Chemistry*, 38: 899–903.

Tayaputh, N. 1996. Present aspects and environmental impacts of pesticide use in Thailand. *Journal of Pesticide Science*, 21: 132–135.

Thao, V.D. 1993. Persistent organochlorine residues in soils from tropical and sub-tropical Asia and Oceania. PhD thesis, Ehime University, Japan, 151pp.

Tolosa, I., L. Merlini, N. de Bertrand, J.M. Bayona, and J. Albaiges. 1992. Occurence and fate of tributyl- and triphenyltin compounds in western Mediterranean coastal enclosures. *Environmental Toxicology and Chemistry*, 11: 145–155.

Tong, S.L., F.Y. Pang, S.M. Phang, and H.C. Lai. 1996. Tributyltin distribution in the coastal environment of peninsular Malaysia. *Environmental Pollution*, No. 91, pp. 209–216.

Toth, S., K. Becker-van slooten, L. Spack, L.F. de Alencastro, and J. Tarradellas. 1996. Irgarol 1051, an antifouling compound in freshwater, sediment, and biota of Lake Geneva. *Bulletin of Environmental Contamination and Toxicology*, No. 57, pp. 426–433.

Uhler, A.D., G.S. Durell, W.G. Steinhaver, and A.M. Spellacy. 1993. Tributyltin levels in bivalve molluscs from the east and west coasts of the United States: Results from the 1988–1990 national status and trends Mussel Watch project. *Environmental Toxicology Chemistry*, 12: 139–153.

Vakily, J.M. 1989. *The Biology and Culture of Mussels of the Genus Perma*. ICLARM Studies and Reviews, No. 17, 63pp., ICLARM, Manila.

Waite, M.E., M.J. Waldock, J.E. Thain, D.J. Smith, and S.M. Milton. 1991. Reductions in TBT concentrations in UK estuaries following legislation in 1986 and 1987. *Marine Environmental Research*, No. 32, pp. 89–111.

Watanabe, S., W. Laovakul, R. Boonyathumanondh, M.S. Tabucanon, and S. Ohgaki. 1996. Concentrations and composition of PCB congeners in the air around stored used capacitors containing PCB insulator oil in a suburb of Bangkok, Thailand. *Environmental Pollution*, No. 92, pp. 289–297.

8

Organochlorine contamination in Baikal seal (*Phoca sibirica*) from Lake Baikal, Russia

Haruhiko Nakata, Shinsuke Tanabe, Ryo Tatsukawa,
Masao Amano, Nobuyuki Miyazaki, and Evgeny A. Petrov

Introduction

Lake Baikal is located in eastern Siberia, Russia (Figure 8.1). It is the deepest fresh water lake in the world (1,632 m) and holds – at a fifth of the world's total – the largest volume of fresh water; it is also known for its geological antiquity (it is over 25 million years old). Among many unique and endemic animals and plants, of most concern is the Baikal seal, a freshwater species. It occupies the highest place in the lake's food chain.

In 1987 and 1988 an acute disease struck the Baikal seal, killing several thousand animals (Grachev *et al.*, 1989). Although the direct cause of these outbreaks was a morbillivirus infection, the factors behind this sudden infection have yet to be understood. Such mass mortalities in marine mammals were frequently found worldwide in the latter half of the twentieth century (Simmonds, 1992). Since these disasters have mostly occurred near industrialized coastal areas, it is suspected that some stresses such as chronic exposure to persistent toxic pollutants such as polychlorinated dibenzo-*p*-dioxins (PCDDs), polychlorinated dibenzofurans (PCDFs), and PCBs (polychlorinated biphenyls) might have played a role in triggering epizootics by suppressing the immune systems of mammals (Kannan *et al.*, 1993; Ross *et al.*, 1996).

Previous investigations reported the presence of high levels of OC residues in marine mammals, and some of them were discussed in association with the occurrence of several abnormalities (Helle, Olsson, and Jensen, 1976; Martineau *et al.*, 1987; Bergman, Olsson, and Reiland, 1992). Adverse effects on reproductive and

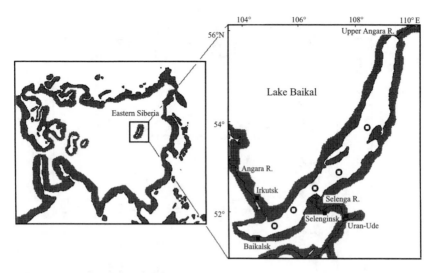

Figure 8.1 Map showing Lake Baikal and the locations (open square in closed circle) where Baikal seals were collected

immunological functions were also noted in captive seals fed with high levels of persistent OCs (Reijnders, 1986; Ross *et al.*, 1996). Nevertheless, unlike marine species, very few investigations are available for freshwater species. To understand the ecotoxicological impacts of persistent organic contaminants on the wildlife, it is necessary to examine their residue levels and construct specific accumulation profiles in various animals.

In this context, the objectives of this study are to understand the status of OC contamination in Baikal seals, specifically in relation to contaminants like DDTs, PCBs, CHLs, and HCHs. Variations in OC levels corresponding with sex and age were also examined in order to elucidate the accumulation pattern. Further, the specific metabolism of PCBs was assessed in Baikal seals, and the 2,3,7,8-TCDD toxic equivalents (TEQ) of non-, mono-, and di-*ortho* coplanar congeners were estimated in order to evaluate this species' potential risk to accumulate highly toxic PCBs.

Status of contamination

Baikal seals were collected from Lake Baikal in May–June, 1992 (Figure 8.1). OCs were detected in all samples of seals and fishes from Lake Baikal (Table 8.1). In Baikal seals, DDT compounds were detected at the highest concentration, ranging from 4.9 to 160 µg/g on a lipid weight basis, followed by PCBs (3.5–64 µg/g), CHLs (0.22–1.9 µg/g), and HCHs (0.028–0.14 µg/g). OC residues in males were significantly higher than those in females ($p < 0.05$). This can be due to the transfer of OCs from female to pup in gestation and lactation as documented for other

Table 8.1 Organochlorine concentrations (mean ± SD, µg/g lipid wt.) in the blubber of Baikal seals and their fish diets (whole body) from Lake Baikal

	Baikal seal		Fish		
	Male	Female	Comephorus baikalensis	Comephorus dybowskii	Concomephorus inermis
N	27	31	8	10	5
Body length (cm)	116 ± 20 (79.5–145.2)	111 ± 15 (79.7–133.2)	15 ± 1.5	12 ± 0.67	16 ± 0.86
Body weight (kg)	51 ± 19 (18.9–90.8)	52 ± 17 (22.1–90.0)	0.020 ± 0.008	0.007 ± 0.002	0.038 ± 0.007
Age (yr)	12 ± 9.2 (0.5–35.5)	9.9 ± 6.2 (0.3–24.5)			
Fat (%)	87 ± 2.8 (81–91)	87 ± 2.3 (80–91)	33	3.8	8.2
p,p'-DDE[†]	30 ± 27 (3.4–110)	10 ± 4.8 (3.1–20)	0.38	0.75	0.19
p,p'-DDD	0.96 ± 0.87 (0.06–4.1)	0.58 ± 0.38 (0.17–1.5)	0.0972	0.11	0.041
p,p'-DDT	15 ± 11 (1.7–50)	9.6 ± 5.8 (1.5–24)	1.3	0.87	0.21
o,p'-DDD	nd (< 0.0016)	nd (< 0.0016)	0.030	0.11	0.058
o,p'-DDT	nd (< 0.0026)	nd (< 0.0026)	0.16	0.18	0.069
Σ-DDT[†]	45 ± 36 (5.3–180)	20 ± 9.9 (4.0–46)	2.0	2.0	0.57
PCBs[†]	22 ± 15 (4.3–64)	11 ± 4.4 (3.5–19)	2.0	3.2	0.82
Oxychlordane[†]	0.51 ± 0.28 (0.12–1.2)	0.24 ± 0.096 (0.11–0.56)	0.006	0.012	0.007
t-nonachlor[†]	0.29 ± 0.14 (0.092–0.62)	0.19 ± 0.064 (0.08–0.32)	0.043	0.10	0.036
c-nonachlor[†]	0.002 ± 0.017 (0.007–0.098)	0.014 ± 0.006 (0.005–0.029)	0.014	0.026	0.010
t-chlordane**	0.002 ± 0.004 (<0.0011–0.01)	0.001 ± 0.002 (<0.0011–0.013)	0.023	0.045	0.019
c-chlordane**	0.009 ± 0.003 (0.004–0.015)	0.009 ± 0.003 (0.005–0.013)	0.040	0.070	0.030
Σ-CHL[†]	0.82 ± 0.43 (0.26–1.9)	0.46 ± 0.15 (0.22–0.83)	0.13	0.25	0.10
α-HCH[†]	0.023 ± 0.008 (0.009–0.039)	0.020 ± 0.008 (0.006–0.045)	0.010	0.010	0.011
β-HCH[†]	0.054 ± 0.024 (0.025–0.11)	0.034 ± 0.014 (0.016–0.089)	0.007	0.006	0.004
γ-HCH	nd (<0.001)	nd (<0.001)	0.001	0.003	0.003
Σ-HCH[†]	0.077 ± 0.027 (0.038–0.14)	0.054 ± 0.020 (0.028–0.11)	0.018	0.019	0.018
DDE/Σ-DDT[†]	0.61 ± 0.14 (0.26–0.86)	0.52 ± 0.12 (0.29–0.72)	0.19	0.38	0.33
oxy/Σ-CHL[†]	0.62 ± 0.06 (0.45–0.70)	0.52 ± 0.06 (0.36–0.67)	0.05	0.05	0.07

Values in parentheses indicate the range.
*Detection limits are designated to be three times the value of blank.
**Concentrations of undetectable samples were regarded as 0 when average and SD were calculated.
[†]Concentrations were significantly different between males and females ($p < 0.05$: Mann-Whitney U-test).

pinnipeds (Addison and Smith, 1974). In fish collected from Lake Baikal, PCBs and DDTs were the dominant components, ranging from 0.82 to 3.2 μg/g and from 0.47 to 2.0 μg/g, respectively (Table 8.1). CHL and HCH levels were about one to three orders of magnitude lower than those of PCBs and DDTs.

OC residue levels found in Baikal seals were compared with those reported in pinnipeds collected from various waters (Table 8.2). The mean concentration of DDTs in adult Baikal seals was 64 μg/g in males and 22 μg/g in females, an order of magnitude higher than those found in seals in the North Sea (Law, Allchin, and Harwood, 1989; Hall *et al.*, 1992), and comparable to harbour seals in the Wadden Sea (Reijnders, 1980) and grey seals in the Baltic Sea (Blomkvist *et al.*, 1992) (Table 8.2). The Baltic and Wadden Seas are known as highly contaminated areas, and previous studies have pointed out the association of the presence of high levels of OCs in seals and disease such as uterine occlusion that result in reproductive failure (Bergman and Olsson, 1985).

Among DDT metabolites, *p,p'*-DDE was the stable compound found in a major proportion of Baikal seals, while *p,p'*-DDT was dominant in fishes (Table 8.1). A ratio of the concentration of *p,p'*-DDE to total DDTs (*p,p'*-DDE/DDTs) can be used to know whether past or present inputs of technical DDT are responsible for the contamination of the ecosystem. The percentage composition of *p,p'*-DDE/DDTs in Baikal seals was 61 ± 14 per cent in males and 52 ± 12 per cent in females, respectively (the average of males and females was 56 ± 14 per cent; Table 8.1). These ratios were lower than those observed for harbour seals affected by a phocine distemper epizootic from 1988 in the North Sea near the UK coast, recording 66 ± 20 per cent (Hall *et al.*, 1992), 66 ± 5.7 per cent (Law, Allchin, and Harwood, 1989), and 64 ± 19 per cent (Mitchell and Kennedy, 1992). In the former USSR, agricultural usage of technical DDT and its production were banned in the 1970s and 1980s, respectively (Barrie *et al.*, 1992). Nevertheless, higher concentrations of DDTs and lower *p,p'*-DDE/DDT ratios in Baikal seals and their fish diet imply a recent input of technical DDT in the watershed of Lake Baikal. Larger variations of DDT compound compositions in air, water, sediment, and soil samples collected around Lake Baikal also indicated the current usage of technical DDTs near the lake (Iwata *et al.*, 1995).

PCB residue levels in adult Baikal seals were 31 μg/g in males and 13 μg/g in females, and about half of these were DDT concentrations (Table 8.2). These were comparable to those reported in harbour seals from the North Sea (Law, Allchin, and Harwood, 1989; Hall *et al.*, 1992) and in grey seals from the Canadian east coast (Addison and Brodie, 1987), but several times lower than those in ringed seals and grey seals from the Baltic Sea (Blomkvist *et al.*, 1992). In Russia, it was estimated that approximately 100,000 and 25,000 tonnes of the technical PCBs, Sovol and trichlorodiphenyls (TCDs), were produced respectively (Ivanov and Sandell, 1992). Relatively higher levels of PCB residues in Baikal seals may suggest its usage or leakage around the Lake Baikal region.

Table 8.2 Comparison of OC concentrations (mean, µg/g lipid wt) in adult Baikal seals with other pinnipeds collected from various waters

Species	Location	Year	N	Age (year)	Sex	Fat (%)	Σ-DDT	Σ-PCB	Σ-CHL	Σ-HCH	References
Baikal seal	Lake Baikal	1992	16	8.5–35.5	M	87	64	31	1.0	0.089	This study
	Lake Baikal	1992	25	8.5–24.5	F	87	22	13	0.47	0.055	This study
Arctic											
Ringed seal	Barrow Strait	1984	19	10.3	M	90	0.79	0.63	0.51	0.30	Muir, Norstrom, and Simon, 1988
	Barrow Strait	1984	14	9.4	F	90	0.53	0.42	0.40	0.34	Muir, Norstrom, and Simon, 1988
Harp seal	Hudson Strait	1989	1	15	M	U	2.3	1.9	na	na	Beck, Smith, and Addison, 1994
	Hudson Strait	1989	1	13	F	U	0.14	0.40	na	na	Beck, Smith, and Addison, 1994
Europe											
Harbour seal	Skagerrak (Sweden)	1988	4	10-22	M	84	7.0	84	na	na	Blomkvist et al., 1992
Grey seal	Baltic Sea	1982–1988	5	9–20	M	90	33	110	na	na	Blomkvist et al., 1992
Ringed seal	Baltic Sea	1981–1986	5	20–40	M	89	340	320	na	na	Blomkvist et al., 1992
Harbour seal	Wadden Sea	1975–1976	8	Adult	U	U	47	700	na	0.40	Reijnders, 1980
	Wash (England)	1989	1	13	M	78	4.6	25	na	na	Hall et al., 1992
	Moray Firth (Scotland)	1989	3	6.7	F	70	3.6	39	na	na	Hall et al., 1992
	West Coast (Scotland)	1989	1	9.0	M	67	2.4	28	na	na	Hall et al., 1992
Grey seal	Blakeney (England)	1988	1	12	F	71	7.1	41	na	0.030	Law, Allchin and Harwood, 1989
	Fahne Island (England)	1988	1	27	F	86	1.2	6.8	na	<0.003	Law, Allchin and Harwood, 1989

	North America										
Harp seal	Newfoundland-Labrador	1970	1	Adult	M	85	50	26	na	na	Frank, Ronald, and Braun, 1973
	Newfoundland-Labrador	1970	13	Adult	F	97	7.1	4.7	na	na	Frank, Ronald, and Braun, 1973
Grey seal	Sable Island (Canada)	1976	6	6.8	F	82	16	14	na	na	Addison and Brodie, 1977
	Sable Island (Canada)	1985	9	10.6	F	94	3.7	30	na	na	Addison and Brodie, 1987
Harp seal	St. Lawrence	1989	13	10.7	M	U	2.1	2.5	na	na	Beck, Smith, and Addison, 1994
	Antartic										
Weddel seal	Showa station	1981	1	13 or 14	M	88	0.19	0.043	na	na	Hidaka, Tanabe, and Tatsukawa, 1983
	Showa station	1981	1	13 or 14	M	88	na	na	0.069	na	Hidaka, Tanabe, and Tatsukawa, 1983

U = unknown; na = no data available

The mean concentrations of CHLs in Baikal seals were 1.0 μg/g for adult males and 0.47 μg/g for adult females, several times lower than those of PCBs and DDTs (Table 8.2). These levels were comparable to ringed seals from the Canadian Arctic (Muir, Norstrom, and Simon, 1988) and harbour seals from Hokkaido (Kawano, 1990) on the northern coast of Japan, both relatively pristine areas. CHL concentrations in Baikal fish (wet weight basis) were also similar to those found in the Canadian Arctic (Muir *et al.*, 1992). Lower levels of CHLs in Baikal seals and fish imply less use of this insecticide near Lake Baikal, as Iwata *et al.* (1995) suggested from abiotic sample data. Oxychlordane and *trans*-nonachlor were the major components of chlordane compounds in Baikal seals, whereas *trans*- and *cis*-chlordane and *trans*- and *cis*-nonachlor were abundant in fishes (Table 8.1).

The mean HCH concentrations were 0.089 μg/g in adult males and 0.055 μg/g in adult females, two to three orders of magnitude lower than those of PCBs and DDTs (Table 8.2). These levels were approximately one order of magnitude lower than those of harp seals and ringed seals in the Canadian Arctic (Muir, Norstrom, and Simon, 1988). Lower HCH concentrations in fish were also observed from Lake Baikal when the levels were compared with those in the Canadian Arctic. In the composition of HCH isomers, α-HCH was dominant in fish, whereas β-HCH was the major component, and γ-HCH was not detected in Baikal seals. Technical HCHs had been extensively used in the former USSR until 1986 (Barrie *et al.*, 1992), and lindane (in a purified form of 90 per cent γ-isomer) is currently being used (Hinckley and Bidleman, 1991). Low concentrations of HCHs, however, as well as non-detectable levels of γ-HCH in Baikal seals suggest the decreasing use of technical HCHs and lindane around Lake Baikal.

Ages and sex-dependent accumulation

OC residue levels varied with age and sex in Baikal seals. In both the sexes, concentrations of PCBs, DDTs, and CHLs increased until maturity (seven to eight years old) (King, 1983). Even after maturity, OC levels increased with age for the males, whereas the levels in females remained constant (Figure 8.2). This pattern is common in cetaceans and pinnipeds, indicating that OCs are accumulated at a faster rate than they are excreted, and that appreciable quantities of OCs are transferred from mother to pup during gestation and lactation (Subramanian, Tanabe, and Tatsukawa, 1987; Aguilar and Borrell, 1988; Tanabe *et al.*, 1994). In the case of HCHs, age-dependent accumulation of residue levels was less pronounced. A similar observation was shown in harbour seals, grey seals (Ofstad and Martinsen, 1983), and northern fur seals (Tanabe *et al.*, 1994), in keeping with the biodegradable and less persistent nature of HCHs.

In grey seals, about 15 per cent of PCB burdens and 30 per cent of DDT burdens in mothers were transferred to pups during the reproductive process

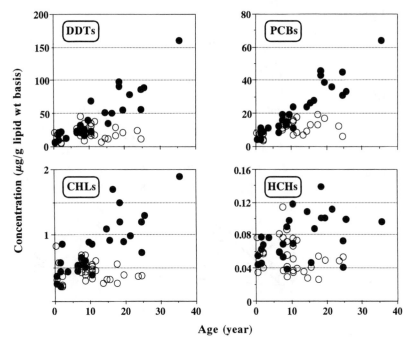

Figure 8.2 Age trends for organochlorine residue levels in male (●) and female (○) Baikal seals

(Addison and Brodie, 1977). To estimate the transfer rates in Baikal seals, the authors calculated PCB and DDT burdens in mature male and females (eight years in males and six years in females; King, 1983). In adult marine mammals, more than 90 per cent of PCB and DDT body burdens were present in the blubber (Hidaka, Tanabe, and Tatsukawa, 1983). Based on this information, the present study estimated burdens on blubber concentrations and the total weight of this tissue measured for each animal at the time of dissection on board. The relationship between age and PCB burden in both sexes is shown in Figure 8.3. The body burdens of PCBs were calculated by the concentrations of PCBs (mg/kg wet weight) in blubber tissues multiplied by the blubber weight (kg). The formulas of regression lines between PCB body burden and age of both sexes of Baikal seals were also calculated, and it was found that body burden = 51 age −130 in males and body burden = 4.4 age + 260 in females (Figure 8.3). An intersecting point obtained from male and female regression lines indicates that in the first reproductive year – 8.4 years old – both male and female seals contained 298 mg of PCBs. It has been reported that 88 per cent of adult female Baikal seals breed every spring (Thomas *et al.*, 1982). Considering this, the body burden of PCBs would be 349 mg in males and 301 mg in females in the next reproductive year. This suggests that 48 mg of PCBs are transferred from mother to pup in every

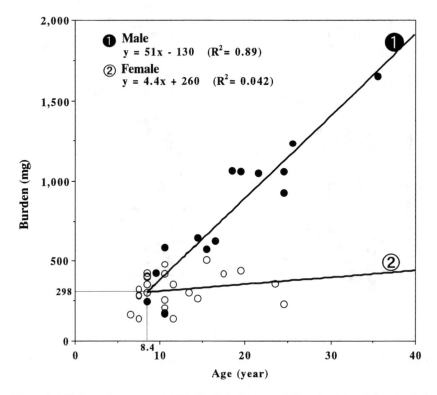

Figure 8.3 Estimated amount of PCBs in Baikal seals, adult males (●) and females (○) transferred from mother to pup

reproductive season. Then, the elimination rate of PCBs was calculated from the following formula:

$$\text{Female burden} = \text{Male burden} \,(1 - 0.01\,P)$$

where P is the transfer rate (percentage). Consequently, 14 per cent of total PCBs in the mother seal are transferred to her pup. When the same approach was used for DDTs, 20 per cent of the transfer rate (135 mg of burden) was obtained. PCB and DDT body burdens in four Baikal seals (six months old) were 62 ± 22 mg and 127 ± 83 mg, respectively. These values were rather close to the transfer quantities from mother to pup (48 mg of PCBs and 135 mg of DDTs). The estimated transfer rates in Baikal seals were comparable to those reported in grey seals (Addison and Brodie, 1977), but apparently lower than those in striped dolphins (Tanabe *et al.*, 1981). The lactation periods of Baikal seals and grey seals are known as 2.5 and 0.5–0.7 months (King, 1983), respectively, whereas that for striped dolphins is 18 months (Miyazaki, 1977). Smaller transfer rates of OCs in seals rather than in dolphins may reflect the shorter lactation period.

PCBs congeners and their metabolism

The mean profile of PCB isomers and congeners in Baikal seals and their fish diets are shown in Figure 8.4. Tri- (3), tetra- (4), and penta- (5) chlorinated congeners in seals were relatively lower than those observed in fishes, indicating that Baikal seals may metabolize some of the lower chlorinated congeners present in their diet. Hexa- (6) chlorinated congeners were dominant in Baikal seals, followed by penta-, hepta-, octa-, and tetrachlorinated homologues. The percentage composition of tetrachlorinated congeners to total PCBs in Baikal seals was 1.5 per cent, considerably lower than that observed for ringed seals in the Canadian Arctic (Muir, Norstrom, and Simon, 1988) and in Ganges River dolphin from India (Kannan, Tanabe, and Tatsukawa, 1994). This fact may suggest that Baikal seals have a higher metabolic capacity to degrade lower chlorinated biphenyls than other aquatic mammals.

The metabolism of PCB isomers and congeners was associated with hepatic mixed-function oxidase such as cytochrome P450 (Safe, 1990; Shimada and Sato, 1980), and PCBs having the vicinal non-chlorinated *meta-para* carbons and *ortho-meta* ones are metabolized by phenobarbital (PB-type) and 3-methyl-cholanthrene (MC-type) induced microsomal enzymes, respectively (Kato, McKinney, and Matthews, 1980; Shimada and Sawabe, 1983; Mills *et al.*, 1985). In order to assess details of the metabolic capacity of Baikal seals, metabolic

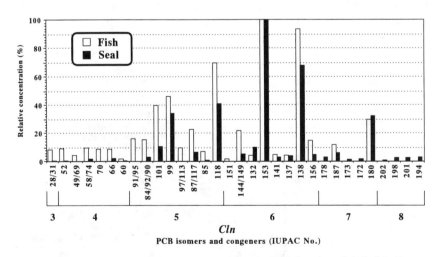

Figure 8.4 PCB isomer and congener compositions in Baikal seals and their fish diet

Note: Relative concentration means the ratio of individual PCB concentrations to that of the maximum peak (IUPAC No. 153), which was treated as 1

index (MI) values of PCB isomers and congeners were calculated following a
model proposed by Tanabe *et al.* (1988):

$$\text{metabolic index (MI}i) = \log \text{CR}180/\text{CR}i$$

where CR*180* is the concentration ratio of PCB isomer No. 180 (IUPAC number)
in fish diets to those in mammals, and CR*i* is the concentration ratio of other isomers. PCB isomers with higher values of MI are more biodegradable than those
with lower MI values. MI values for specific PCB isomers in different species of
higher animals (Tanabe *et al.*, 1988) were compared (Figure 8.5). Interestingly,
MI values for PCB isomers metabolized by PB-type enzymes (PCB No. 52) were
higher in Baikal seals compared to other marine mammals, but apparently lower
than in terrestrial mammals. On the other hand, MI values for PCB isomers
metabolized by MC-type enzymes (PCB No. 66) were lower in Baikal seals, and
comparable to those in cetaceans. Earlier studies demonstrated that cetaceans
have a smaller capacity to degrade toxic contaminants, due to poorly functioning
PB- and MC-type enzymes. Cetaceans are, therefore, recognized as receiving
high concentrations of persistent OCs (Tanabe *et al.*, 1988; Watanabe *et al.*,
1989). Baikal seals have a higher capacity to degrade OCs than cetaceans, but on
the whole still lower than other terrestrial mammals (Figure 8.5). This may be
linked to the high accumulation nature of OC residue levels in Baikal seals.
Comparing MI values among pinnipeds, Baikal seals were estimated to have
more active PB-type enzymes than other marine species, and apparently less
active MC-type enzymes (Figure 8.5). These differences in MI values may imply
the different types of drug-metabolizing enzyme systems in the degradation of
xenobiotics and also suggest the occurrence of different types of toxic effects
between seals inhabiting fresh water and marine water.

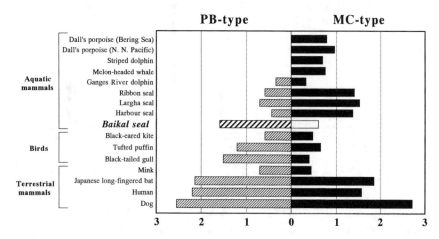

Figure 8.5 Estimated PB-type and MC-type enzyme activities in higher trophic animals
by metabolic indices of 2,2′,5,5′- (PB-type) and 2,3′,4,4′- (MC-type) tetrachlorobiphenyls

Toxicological implications

The estimated 2,3,7,8-TCDD toxic equivalents of non-, mono-, and di-*ortho* coplanar congeners in Baikal seals are shown in Table 8.3 and compared with those of marine mammals (Table 8.4). The mean TCDD equivalents of coplanar congeners was 2,500 ± 2,100 pg/g on a wet weight basis, apparently lower than those of diseased/stranded striped dolphin, bottlenose dolphin (Kannan *et al.*, 1993), and Risso's dolphin in the Mediterranean Sea (Corsolini *et al.*, 1995), but higher than those of harbour seals in the North Sea (Storr-Hansen and Spliid, 1993) and ringed seals in Spitsbergen in the Norwegian Arctic (Daelemans *et al.*, 1993). The concentration ratio of total TEQs to total PCBs in Baikal seals was, however, 20×10^{-5}, approximately 10 times higher than that of the harbour seal and three to six times higher than those of cetaceans, comparable to those of ringed seals. These results indicate that the contribution of toxic PCB congeners to total TEQs is more prominent in Baikal seals and ringed seals than in other marine mammals, with the possible reason being the species-specific capacity for biotransformation of such congeners. In order to understand the relative persistence of non-, mono-, and di-*ortho* coplanar congeners in aquatic mammals, the ratios of TEQ of individual coplanar congeners to that of IUPAC 153 were calculated and are presented in Table 8.4. Since IUPAC 153 is one of the most persistent congeners in biota, the congeners with higher ratios imply a relatively more persistent nature and contribute significantly to the total TEQs. Interestingly, the ratios of mono-*ortho* congeners in Baikal seals and ringed seals, particularly IUPAC 105 and 118, were about 2.5 to 10 times higher than those in other marine mammals. In addition, the ratios of non-*ortho* congeners in this species, such as IUPAC 77 and 126, were also higher than those in other mammals.

Considering the foregoing observation, it is plausible to assume that even if the total PCB concentration in Baikal seals is comparable to those in other mammals, the risk of TCDD remains serious because of the high accumulation of toxic PCB congeners. Among the wide range of toxicities of TCDD-related compounds, the adverse effects caused by enzymatic induction and immunological dysfunction are of greatest concern. Recently it was found that chronic exposure to organochlorine contaminants, particularly to toxic PCB congeners, is responsible for immunosuppression in the harbour seal, which might have triggered the virus-induced mass mortalities in the North Sea in 1988 (Swart *et al.*, 1994; Ross *et al.*, 1996). Further studies are needed in terms of the relationship between the enrichment of toxic PCB congeners and possible toxicities, in particular immunosuppression, since mass mortality of seals by morbillivirus infection also took place in Lake Baikal in 1987–1988.

Table 8.3 Concentrations of di-, mono-, and non-*ortho* coplanar congeners and total PCBs (ng/g wet wt), and their 2,3,7,8-TCDD toxic equivalents (pg/g) in the blubber of Baikal seals

	n	Age (years)	Body length (cm)	Body weight (kg)	Lipid (%)	77	non-*ortho* 126	169
Male seal								
Immature	7	1.6 ± 1.0 (0.5–3.5)	104 ± 6.8 (95–110)	29 ± 10 (19–48)	86 ± 2.8 (82–90)	6.3 ± 4.0 (2.5–14)	2.4 ± 1.3 (0.6–4.7)	0.19 ± 0.10 (0.02–0.3)
Mature	8	16 ± 10 (7.5–35.5)	145 ± 7.7 (137–156)	63 ± 13 (49–90)	89 ± 2.7 (86–94)	23 ± 9.8 (12–41)	3.5 ± 1.2 (2.3–5.9)	0.37 ± 0.19 (0.20–0.70)
Female seal								
Immature	6	1.3 ± 0.8 (0.3–2.0)	102 ± 7.3 (94–111)	27 ± 5.0 (22–34)	87 ± 1.6 (85–89)	8.8 ± 7.6 (3.1–23)	2.5 ± 2.6 (1.0–7.7)	0.30 ± 0.10 (0.20–0.40)
Mature	19	11 ± 4.9 (6.5–24.5)	138 ± 6.3 (125–149)	57 ± 14 (38–90)	87 ± 1.9 (83–91)	14 ± 8.1 (5.6–33)	3.4 ± 2.0 (1.0–9.0)	0.30 ± 0.20 (0.06–0.8)
Fish*	23	—	14 ± 1.9 (11–17)	0.018 ± 0.013 (0.005–0.049)	15 ± 15 (3.8–33)	1.0 ± 1.2 (0.2–2.4)	0.30 ± 0.30 (0.09–0.6)	0.07 ± 0.08 (0.01–0.20)
TEQ (seal)						140 ± 94	310 ± 190	14–8.5

Male seal							
Immature	250 ± 130	650 ± 370	87 ± 57	180 ± 110	650 ± 360	990 ± 580	250 ± 190	5,000 ± 2,700
	(65–410)	(180 ± 1,200)	(26–120)	(46–350)	(180–1,100)	(260–2,000)	(53–610)	(1,400–9,500)
Mature	1,200 ± 830	2,800 ± 1,800	260 ± 170	1,300 ± 720	3,800 ± 2,900	6,700 ± 4,500	1,700 ± 1,300	26,000 ± 17,000
	(420–2,700)	(1,100–5,800)	(130–620)	(600–2,500)	(1,300–9,900)	(2,800–15,000)	(770–4,200)	(11,000–59,000)
Female seal							
Immature	260 ± 160	740 ± 470	74 ± 41	160 ± 100	650 ± 470	990 ± 740	190 ± 140	5,000 ± 3,200
	(140–560)	(420–1,700)	(41–140)	(84–350)	(310–1,500)	(460–2,400)	(93–460)	(3,000–11,000)
Mature	360 ± 160	960 ± 500	150 ± 67	370 ± 190	1,400 ± 560	2,400 ± 880	710 ± 280	10,000 ± 3,800
	(110–650)	(61–1,700)	(28–300)	(70–720)	(540–2,500)	(860–4,000)	(230–1,300)	(4,100–17,000)
Fish	12 ± 13	30 ± 30	5.1 ± 4.7	5.7 ± 4.0	32 ± 32	34 ± 32	11 ± 9.1	350 ± 350
	(8.3–65)	(3.4–27)	(1.6–11)	(2.3–10)	(9.1–68)	(9.7–70)	(3.9–21)	(98–750)
TEQ (seal)	490 ± 510	1,200 ± 1,200	150 ± 110	9.1 ± 11	33 ± 35	56 ± 58	15 ± 16	2,500 ± 2,100

TEF values cited from Safe (1990). Values it parenthesis is the range. Detection limits <0.003 ng/g on weight basis.
*Three species of fish were involved

Table 8.4 Concentration ratios, total TEQs/total PCBs, and coplanar congeners TEQs/IUPAC 153 TEQs in aquatic mammals

	Baikal seal	Ringed seal[1]	Harbour seal[2]	Bottlenose dolphin[3]	Striped dolphin[4]	Risso's dolphin[3]
Total TEQs (ng/g)	2.5	0.19	0.47	19	18	21
Total PCBs (ng/g)	12,000	1,400	26,000	584,000	343,000	320,000
TEQ/PCBs (×10)	20	13	1.8	3.1	4.6	6.4
TEQ ratio to #153						
77*	2.5	0.18	0.04	0.10	0.29	0.12
126*	5.5	3.4	0.42	0.12	0.47	0.32
169*	0.25	NA	0.017	0.12	0.27	0.22
105**	8.8	7.2	0.65	0.73	1.4	2.8
118**	21	19	1.8	2.4	5.4	8.0
156**	2.7	1.6	0.80	0.80	2.1	1.6
137†	0.16	0.020	0.009	0.035	0.075	0.045
138†	0.59	0.67	0.72	0.72	0.82	0.73
153†	1	1	1	1	1	1
180†	0.27	0.15	0.18	0.38	0.53	0.55

NA: no data available
*non-ortho
**mono-ortho
†di-ortho congeners
[1] Daeleman et al. (1993)
[2] Storr-Hensen and Spliid (1993)
[3] Corsolini et al. (1995)
[4] Kannan et al. (1993)

Conclusions

Persistent OCs such as DDTs, PCBs, CHLs, and HCHs were determined in Baikal seals and their fish diets. Although OC contamination in air, water, and sediments is not so prominent in global terms (Iwata *et al.*, 1995), noticeably high concentrations of DDTs and PCBs were detected in Baikal seals – concentrations comparable to those in some species of pinnipeds from polluted waters, such as the North Sea, Baltic Sea, and Canadian east coast. In this context, it can be concluded that the problem of OC pollution in Lake Baikal rests on seals that retain OCs with unexpectedly high residue levels, even while the water remains relatively clean. Based on the remarkable accumulation of persistent OCs in Baikal seals attributable to their lower drug-metabolizing capacity and to their reproductive transfer, long-term contamination and concomitant toxic effects over generations are of great concern in this region. Although the decisive connection between high levels of OC accumulation and mass mortality in Baikal seal is still unclear, the levels of OC residues may disturb the function of cytochrome P450 monooxygenase and depress the immune system in these animals. Additional research on OC monitoring, including dioxins and related compounds in association with drug-metabolizing enzyme systems, is needed to understand the comprehensive toxic effects on Baikal seals.

REFERENCES

Addison, R.F. and T.G. Smith. 1974. Organochlorine residue levels in Arctic ringed seals: variation with age and sex. *OIKOS*, 25: 335–337.

Addison, R.F. and P.F. Brodie. 1977. Organochlorine residues in material blubber, milk, and pup blubber from grey seals (*Halichoerus grypus*) from Sable Island, Nova Scotia. *Journal of the Fisheries Research Board of Canada*, 34: 937–939.

Addison, R.F. and P.F. Brodie. 1987. Transfer of organochlorine residues from blubber through the circulatory system to milk in the lactating grey seal *Halichoerus grypus*. *Canadian Journal of Fisheries and Aquatic Sciences*, 44: 782–786.

Aguilar, A. and A. Borrell. 1988. Age- and sex-related changes in organochlorine compound levels in Fin whales *(Balaenoptera physalus)* from the Eastern North Atlantic. *Marine Environment Research*, 25: 195–211.

Barrie, L.A., D. Gregor, H. Hargrave, R. Lake, D. Muir, R. Shearer, B. Tracey, and T. Bidlemam. 1992. Arctic contaminants: sources, occurrence and pathways. *The Science of the Total Environment*, No. 122, 1–74.

Beck, G.G., T.G. Smith, R.F. Addison. 1994. Organochlorine residues in harp seals, *Phoca groenlandica*, from the Gulf of St. Lawrence and Hudson Strait: an evaluation of contaminant concentrations and burdens, *Canadian Journal of Zoology*, No. 72, 174–182.

Bergman, A. and M. Olsson. 1985. Pathology of Baltic grey seal and ringed seal females with special reference to adrenocortical hyperplasia: Is environmental pollution the cause of a widely disturbed disease syndrome? *Finnish Game Research*, 44: 47–62.

Bergman, A., M. Olsson, and S. Reiland. 1992. Skull-bone lesions in the Baltic grey seal (*Halichoerus grypus*), *AMBIO*, 21: 517–519.

Blomkvist, G., A. Roos, S. Jensen, A. Bignert, and M. Olsson. 1992. Concentration of sDDT and PCB in seals from Swedish and Scottish waters. *AMBIO*, 8: 539–545.

Corsolini, S., S. Focardi, K. Kannan, S. Tanabe, A. Borrell, and R. Tatsukawa. 1995. Congener profile and toxicity assessment of polychlorinated biphenyls in dolphins, sharks and tuna collected from Italian coastal waters. *Marine Environment Research*, 40: 33–53.

Daelemans, F.F., F. Mehlum, C. Lydersen, and P.J.C. Schepens. 1993. Mono-ortho and non-ortho substituted PCBs in Arctic ringed seal (*Phoca hispida*) from the Svalbard area: analysis and determination of their toxic threat. *Chemosphere*, 27: 429–437.

Frank, R., K. Ronald, H.E. Braun. 1973. Organochlorine residues in harp seals (*Pagophilus groenlandicus*) caught in Eastern Canadian waters. *Journal of Fisheries Research Board of Canada*, 30: 1053–1063.

Grachev, M.A., V.P. Kumarev, L.V. Mamaev, V.L. Zorin, L.V. Baranova, N.N. Denikina, S.I. Belikov, E.A. Petrov, V.S. Kolesnik, R.S. Kolesnik, V.M. Dorofeev, A.M. Beim, V.N. Kudelin, F.G. Nagieva, and V.N. Sidorov. 1989. Distemper virus in Baikal seals. *Nature*, 38: 209.

Hall, A.J. R.J. Law, D.E. Wells, J. Harwood, H.M. Ross, S. Kennedy, C.R. Allchin, L.A. Campbell, and P.P. Pomeroy. 1992. Organochlorine levels in common seals (*Phoca vitulina*) which were victims and survivors of the 1988 phocine distemper epizootic. *Science of the Total Environment*, No. 115, 145–162.

Helle, E., M. Olsson, and S. Jensen. 1976. PCB levels correlated with pathological changes in seal uteri. *AMBIO*, 5: 261–263.

Hidaka, H., S. Tanabe, and R. Tatsukawa. 1983. DDT compounds and PCB isomers and congeners in Weddell seals and their fate in the Antarctic marine ecosystem. *Agricultural and Biological Chemistry*, 47: 2009–2017.

Hinckley, D.A. and T. Bidleman. 1991. Atmospheric organochlorine pollutants and air-sea exchange of hexachlorocyclohexane in the Bering and Chukchi Seas. *Journal Geophysical Research*, 96: 1702–1713.

Ivanov, V. and E. Sandell. 1992. Characterization of polychlorinated biphenyl isomers in Sovol and trichlorodiphenyl formulations by high-resolution gas chromatography with electron capture detection and high-resolution gas chromatography-mass spectrometry techniques. *Environmental Science and Technology*, 26: 2012–2017.

Iwata, H., S. Tanabe, K. Ueda, and R. Tatsukawa. 1995. Persistent organochlorine residues in air, water, sediments and soils from Lake Baikal regions, Russia. *Environmental Science and Technology*, 29: 792–801.

Kannan, K., S. Tanabe, A. Borrell, A. Aguilar, S. Focardi, and R. Tatsukawa. 1993. Isomer-specific analysis and toxic evaluation of polychlorinated biphenyls in striped dolphins affected by an epizootic in the Western Mediterranean Sea. *Archives of Environment Contamination and Toxicology*, 25: 227–233.

Kannan, K., S. Tanabe, and R. Tatsukawa. 1994. Biodegradation capacity and residue pattern of organochlorines in Ganges river dolphin. *Toxicological and Environment Chemistry*, No. 42, 249–261.

Kato, S., J.D. McKinney, and H.B. Matthews. 1980. Metabolism of symmetrical hexachloro-biphenyl isomers in the rat. *Toxicology and Applied Pharmacology*, No. 53, 389–398.

Kawano, M. 1990. PhD thesis, Kyushu University (in Japanese).

King, J.E. 1983. *Seal of the World*, London: British Museum.

Law, R.J., C.R. Allchin, and J. Harwood. 1989. Concentrations of organochlorine compounds in the blubber of seals from eastern and north-eastern England, 1988. *Marine
Pollution Bulletin*, 20: 110–115.

Martineau, D., P. Béland, C. Desjardins, and A. Lagacé. 1987. Levels of organochlorine
chemicals in tissues of Beluga whales (*Delphinapterus leucas*) from the St. Lawrence
Estuary, Quebec, Canada. *Archives of Environment Contamination and Toxicology*, 16:
137–147.

Mills, R.A., C.D. Millis, G.A. Dannan, F.P. Guengerich, and S.D. Aust. 1985. Studies on
the structure-activities relationships for the metabolism of polychlorinated biphenyls by
rat liver microsomes. *Toxicology and Applied Pharmacology*, 78: 96–104.

Mitchell, S.H. and S. Kennedy. 1992. Tissue concentrations of organochlorine compounds
in common seals from the coast of Northern Ireland. *Science of the Total Environment*,
115: 163–177.

Miyazaki, N. 1977. Growth and reproduction of *Stenella coeruleoalba* off the pacific coast
of Japan. *Scientific Reports of the Whales Research Institute*, No. 29, 21–48.

Muir, D.C.G., R.J. Norstrom, and M. Simon. 1988. Organochlorine contaminants in arctic
marine food chain: Accumulation of specific polychlorinated biphenyls and chlordane-
related compounds. *Environmental Science and Technology*, 22: 1071–1079.

Muir, D.C.G., R. Wagemann, B.T. Hargrave, D.J. Thomas, D.B. Peakall, and
R.J. Norstrom. 1992. Arctic marine ecosystem contamination. *Science of the Total
Environment*, 122: 75–134.

Ofstad, E.B. and K. Martinsen. 1983. Persistent organochlorine compounds in seals from
Norwegian coastal waters. *AMBIO*, 12: 262–264.

Reijnders, P.J.H. 1980. Organochlorine and heavy metal residues in harbour seals from the
wadden sea and their possible effects on reproduction. *Netherlands Journal of Sea
Research*, No. 14, 30–65.

Reijnders, P.J.H. 1986. Reproductive failure in common seals feeding on fish from
polluted coastal waters. *Nature*, No. 324, 456–457.

Ross, P., R.D. Swart, R. Addison, H.V. Loveren, J. Vos, and A. Osterhaus. 1996.
Contaminant-induced immunotoxicology in harbour seals: wildlife at risk? *Toxicology*,
No. 112, 157–169.

Safe, S. 1990. Polychlorinated biphenyls (PCBs), dibezo-*p*-dioxins (PCDDs), dibenzofurans (PCDFs), and related compounds: Environmental and mechanistic considerations
which support the development of toxic equivalency factors. *CRC Critical Reviews in
Toxicology*, 21: 51–88.

Shimada, T. and Y. Sawabe. 1983. Activation of 3,4,3',4'-tetrachlorobiphenyl to protein-
bound metabolites by rat liver microsomal cytochrome P-448-containing monooxygenase system, *Toxicology and Applied Pharmacology*, No. 70, 486–493.

Shimada, T and R. Sato, 1980. Covalent binding of polychlorinated biphenyls to rat liver
microsomes *in vitro* nature of reactive metabolites and target macromolecules,
Toxicology and Applied Pharmacology, No. 55, 490–500.

Simmonds, M. 1992. Green Peace International Mediterranean Sea Project, Mediterranean
striped dolphin mortality international symposium, Madrid, Spain.

Storr-Hansen, E. and H. Spliid. 1993. Distribution patterns of polychlorinated biphenyls
congeners in harbor seal (*Phoca vitulina*) tissues: statistical analysis. *Archives of
Environment Contamination and Toxicology*, No. 25, 328–345.

Subramanian, A., S. Tanabe, and R. Tatsukawa. 1987. Age and size trends and male-female
differences of PCBs and DDE in dall-type Dall's porpoises, *Phocoenoides dalli* of

north-western North Pacific. *Proceedings of the NIPR Symposium on Polar Biology*, 1: 205–216.

Swart, R.L.D., P.S. Ross, L.J. Vesser, H.H. Timmerman, S. Heisterkamp, H.V. Loveren, J.G. Vos, P.J.H. Reijnders, and A.D.M.E. Osterhaus. 1994. Impairment of immune function in harbour seals (*Phoca viturina*) feeding on fish from polluted waters. *AMBIO*, 23: 155–159.

Tanabe, S., H. Tanaka, K. Maruyama, R. Tatsukawa. 1981. Elimination and chlorinated hydrocarbons from mother striped dolphin (*Stenella coeruleoalba*) through parturition and lactation. In: *Studies on the Levels of Organochlorine Compounds and Heavy Metals in the Marine Environment*, T. Fujiyama (ed), pp. 115–121. University of Ryukyus, Ryukyus.

Tanabe, S., S. Watanabe, H. Kan, and R. Tatsukawa. 1988. Capacity and mode of PCB metabolism in small cetaceans. *Marine Mammal Science*, 4: 103–124.

Tanabe, S., J.-K. Sung, D.-Y. Choi, N. Baba, M. Kiyota, K. Yoshida, and R. Tatsukawa. 1994. Persistent organochlorine residues in northern fur seal from the Pacific coast of Japan since 1971. *Environment Pollution*, No. 85: 305–314.

Thomas, J., V. Pastukhov, R. Elsner, and E. Petrov. 1982. *Phoca sibirica, Mammalian Species*, No. 188, 1–6.

Watanabe, S., T. Shimada, S. Nakamura, N. Nishiyama, N. Yamashita, S. Tanabe, and R. Tatsukawa. 1989. Specific profile of liver microsomal cytochrome P-450 in dolphin and whale. *Marine Environment Research*, No. 27, 51–65.

9

Marine mammals and environmental contaminants in the Pacific Ocean: Current knowledge and frontiers for research

Thomas J. O'Shea

Introduction

Concern about the impact of environmental contaminants on marine mammals has increased greatly. Over the last decade the number of compounds reported in tissues of marine mammals globally exceeded 200 and the number of individual animals sampled for contaminant residues was more than double the historical total since analyses began in the 1960s (O'Shea and Tanabe, 2003). This concern has been heightened by experiments and observations that implicate contaminants as detrimental to marine mammal health and reproduction. However, the state of our understanding of the degree and implications of contamination of marine mammals is very incomplete, particularly from a large-scale perspective. The objectives of this chapter are therefore twofold. First, it provides an overview of current knowledge about persistent environmental contaminants in marine mammals of the Pacific Ocean. This is followed by a synopsis of results of a workshop held in October 1998 at Keystone, Colorado, USA (O'Shea, Reeves, and Kirk Long, 1999). At that workshop about 60 experts developed statements describing the most important scientific questions and topics in this field, or the "Frontiers for Research". These apply equally to marine mammals in the Pacific Ocean or anywhere else in the world.

Marine mammals and persistent contaminants in the Pacific Ocean

Current knowledge

Broad patterns in current knowledge about persistent contaminants in marine mammals of the Pacific Ocean can be summarized based on a review of more than 400 publications on marine mammals and contaminants worldwide that have appeared through 1998, primarily in scientific journals. Full citations to these references are available in three recent overview papers (Fossi and Marsili, 1997; O'Shea, 1999; O'Shea and Tanabe, 2003). Although these reviews have undoubtedly missed some information, they provide the basis for the summaries and compilations provided in this chapter. They probably encompass most of what has been published on this topic through to the end of the International Year of the Ocean. These summaries are at the least highly representative of the emphasis of the work in this field up until this time. This overview does not discuss the effects of oil on marine mammals, although based on the *Exxon Valdez* accident in Alaska, some of the most detailed field studies of the impact of oil on marine mammals come from the Pacific Ocean. The reader should consult sources in the volumes edited by Loughlin (1994) and Geraci and St Aubin (1990) for summaries of knowledge on the impact of oil on marine mammals.

Contaminant surveys in Pacific Ocean marine mammals

Globally, the study of contaminants in marine mammals has emphasized surveys of organochlorine compounds and potentially toxic elements in tissues. Organochlorine compounds can be classified on the basis of their chemical structures, their sources, and their original intended uses. All are carbon-based (organic) compounds that contain various numbers of chlorine atoms. The structure of these organic molecules can be relatively simple, or can involve ring structures and other complex configurations. In many cases their chemical properties make them resistant to metabolism, and they differentially absorb in fatty tissues. These two properties allow them to persist in organisms and to concentrate in food webs, reaching highest concentrations in fatty tissues of consumers at the tops of these food webs (biomagnification). Many fish-eating marine mammals are high-level consumers with large amounts of fat stored in their blubber. In some cases the original organochlorine chemical is less persistent than the metabolites that are produced by living systems. Many of the well-known and often-measured organochlorines originate from pesticides, particularly insecticides. DDT (p,p'-DDT or 2,2-bis-(p-chlorophenyl)-1,1,1-trichloroethane) is a well-known example. It is an insecticide currently banned from use in many countries because of its toxicity, and because of the persistence and biomagnification of its principal metabolite (p,p'-DDE (2,2-bis-(p-chlorophenyl)-1,1,1-dichloroethylene), which has been detected in blubber of every marine mammal examined. Other organochlorine insecticides or their metabolites commonly reported in analyses of marine mammal tissues include a group called cyclodienes. These include dieldrin, used as an insecticide and also produced as a metabolite of the insecticide aldrin, and endrin. A somewhat structurally related mixture of insecticidal compounds known as chlordanes and metabolites (e.g. heptachlor epoxide) are found in marine mammals, as are the insecticides mirex, toxaphene (a complex mixture of chlorinated compounds), hexachlorocyclohexanes (HCHs), and the fungicide hexachlorobenzene (HCB).

Some organochlorine contaminants originate as compounds used for industrial purposes. These include the polychlorinated biphenyls (PCBs), which occur as complex mixtures of potentially over 200 different compounds (referred to as individual PCB congeners) that include two organic ring structures (biphenyls) and variable numbers of associated chlorine atoms. These are ubiquitous in the global marine environment and have been associated with a variety of harmful effects on animals. They were originally produced for uses in plastics, hydraulic fluids, electrical transformers, and many other applications. Although production has been drastically reduced, PCBs continue to enter the environment from past applications and existing uses, and many scientists view them as very serious contaminants because of their persistence and toxicity. Other industrially based or non-insecticidal organochlorines can be found in marine mammals, including the highly toxic polychlorinated dibenzodioxins (PCDDs), dibenzofurans (PCDs), and other compounds noted elsewhere in this chapter.

Chemical elements are often examined in tissues of marine mammals, particularly in samples taken from livers and kidneys. These are naturally occurring, and by definition cannot be further broken down chemically and thus persist forever. Many elements are essential to the function of living systems and are not harmful except at high concentrations. However, human activities have increased the availability of many elements in the environment, and some elements have no known biological role and can be toxic at low amounts. Mercury, cadmium, and lead are well known in this regard and are often investigated in studies of marine mammals. Some elements can be assimilated into organic forms that are much more toxic than the simple element, as is well known for methyl mercury. Organic complexes with tin, particularly butyltin compounds, are produced industrially (particularly as marine paints) and have recently been found in tissues of marine mammals (see later). Radionuclides are persistent radioactive by-products of fallout from nuclear weapons testing, nuclear reactor accidents and operations, nuclear fuels, and other technical applications.

The Pacific Ocean is not as well represented in sampling of marine mammals for contaminants as the Arctic or Atlantic Oceans. The references noted in Table 9.1 provide information on 19,924 individual marine mammals examined for organochlorines and toxic elements worldwide, but only 14.5 per cent (about 2,882) of these have been from the Pacific Ocean. Approximately 80 species of marine mammals currently inhabit the Pacific Ocean and its associated seas (Mead and Brownell, 1993). These include roughly 10 species of mysticete

Table 9.1 Summary of numbers of individual marine mammals sampled for organochlorine or metals analysis in the Pacific Ocean

Group	Organochlorines	Metals	Total
Pinnipeds	828	673	1,501
Odontocete cetaceans	513	520	1,033
Mysticete cetaceans	85	26	111
Dugongs	2	47	49
Sea otters	70	58	128
Total	1,498	1,324	2,822

Note: Totals are based on sources published up until the end of 1998 and cited in literature reviews compiled by Fossi and Marsili (1997), O'Shea (1999), and O'Shea and Tanabe (2003)

cetaceans (baleen whales), 47 species of odontocetes (toothed whales, porpoises, and dolphins), 20 species of pinnipeds (seals, sea-lions, and walrus), the dugong (*Dugong dugon*), and the sea otter (*Enydra lutris*). Information about the occurrence of some contaminants in individuals sampled from the Pacific Ocean is available for about 60 per cent (48) of these species (Tables 9.2, 9.3, and 9.4). Approximately equal numbers of individuals have been sampled for organochlorines as for toxic elements across all marine mammals (Table 9.1), but in many species there are very few data for these contaminants. Sample sizes of less than 20 individuals for either organochlorines or elements characterize all but nine of these species (more than 80 per cent).

More pinnipeds have been sampled for environmental contaminants than odontocetes; small numbers of mysticetes, dugongs, and sea otters have been sampled (Table 9.1). These totals are cumulative from about 1970 (when the first publications on the topic appeared for the Pacific Ocean) through to the end of 1998. Sampling of marine mammals for contaminants in the Pacific Ocean has been geographically biased. Only about 7.5 per cent of the samples are from the southern hemisphere, and most of these are from eastern Australia and New Zealand. Nearly 85 per cent of the individuals come from three broad areas: 36 per cent are from the north-west Pacific near the coast of Asia; 17 per cent are from the Bering Sea and Pacific coast of Alaska; and 31 per cent are from British Columbia and the west coast of the USA (Figure 9.1). Thus most knowledge about contaminants in marine mammals in the Pacific Ocean pertains to results of sampling from these three regions.

Despite the smaller numbers of samples and restricted sampling regions, marine mammals in the Pacific Ocean have been reported to contain the same array of organochlorine contaminants in their blubber as marine mammals from other oceans. Organochlorine contaminants found in blubber include pesticides and their breakdown products such as DDT and metabolites (specifically p,p'-DDE, p,p'-DDD, and their o,p isomers as well as methyl sulfone metabolites);

Table 9.2 Summaries of number of individual pinnipeds sampled for organochlorine and metal contamination in the Pacific Ocean, based on publications upto the end of 1998

Species	Organochlorines	Metals	Principal references
Arctocephalus phillipi (Juan Fernández fur seal)	—	29	Sepúlveda, Ochoa-Acuña, and Sundlof, 1997
Callorhinus ursinus (northern fur seal)	322	196	Anas and Wilson, 1970; Anas, 1971, 1973, 1974a; Kim, Chu, and Barron, 1974; Anas and Worlund, 1975; Kurtz and Kim, 1976; Bacon Jarmion, and Costa, 1992; Britt and Howard, 1983; Goldblatt and Anthony, 1983; Richard and Skoch, 1986; Mössner et al., 1992; Varanasi et al, 1992, 1993b; Schantz et al., 1993; Zeisler et al., 1993; Tanabe et al., 1994; Noda et al., 1995; Becker et al., 1997; Krahn et al., 1997; Mössner and Ballschmiter, 1997; Iwata et al., 1998; Tanabe et al., 1998
Erignathus barbatus (bearded seal)	6	—	Krahn et al., 1997
Eumetopias jubatus (Steller's sea-lion)	56	137	Hamanaka, Itoo, and Mishima, 1982; Varanasi et al., 1992, 1993b; Tanabe et al., 1994; Kim et al., 1996a, 1996c; Lee et al., 1996; Hayteas and Duffield, 1997; Sydeman and Jarman, 1998
Histriophoca fasciata (ribbon seal)	—	29	Hamanaka, Kato, and Tsujita, 1977; Tanabe et al., 1998
Hydrurga leptonyx (leopard seal)	40	1	Kemper et al., 1994
Mirounga angustirostris (northern elephant seal)	40	—	Britt and Howard, 1983; Schafer et al., 1984; Bacon, Jarmon, and Costa, 1992; Newman et al., 1994; Hayteas and Duffield, 1997; Beckmen et al., 1997
Monachus schauinslandii (Hawaiian monk seal)	1	—	Takei and Leong, 1981
Odobenus rosmarus (walrus)	53	121	Taylor, Schliebe, and Metsker, 1989; Warburton and Seagars, 1993
Phoca largha (spotted seal)	4	3	Serat et al., 1977; Iwata et al., 1994, 1997; Tanabe et al., 1994, 1998
Phoca vitulina (harbour seal or common seal)	110	70	Shaw, 1971; Anas, 1974a, 1974b; Hamanaka, Kato, and Tsujita, 1977; Britt and Howard, 1983; Tohyama et al., 1986; Himeno et al., 1989; Walker et al., 1989; Varanasi et al., 1992, 1993b; Hayteas and Duffield, 1997; Krahn et al., 1997; Young et al., 1998
Pusa hispida (ribbon seal; also referred to as *Phoca hispida*)	8	14	Hamanaka, Kato, and Tsujita, 1977; Mackey et al., 1996; Krahn et al., 1997
Zalophus californianus (California sea-lion)	228	73	Anas, 1971; LeBouef and Bonnell, 1971; Hall, Gilmartin, and Mottson, 1971; Braham, 1973; DeLong, Gilmartin, and Simpson, 1973; Theobald, 1973; Buhler, Claeys, and Mate, 1975; Martin et al., 1976; Lee et al., 1977; Britt and Howard, 1983; Bacon, Jarmon, and Costa, 1992; Lieberg-Clark et al., 1995; Hayteas and Duffield, 1997

Note: Organotin determinations are included with metals. Scientific names are given with common names in parentheses, following the nomenclature of Rice (1998)

115

Table 9.3 Summaries of number of individual odontocete cetaceans sampled for organochlorine and metal contamination in the Pacific Ocean, based on publications up until the end of 1998

Species	Organochlorines	Metals	Principal references
Berardius bairdii (Baird's beaked whale)	49	—	Subramanian, Tanabe, and Tatsukawa, 1988a; Prudente et al., 1996, 1998; Kannan et al., 1989
Delphinus delphis (short-beaked common dolphin)	33	5	Koeman et al., 1972, 1975; O'Shea et al., 1980; Iwata et al., 1994; Kemper et al., 1994; Prudente et al., 1997; Hayteas and Duffield, 1997; Tanabe et al., 1996; Schafer et al., 1984
Delphinapterus leucas (beluga whale)	—	2	Zeisler et al., 1993; Becker et al., 1995
Globicephala macrorhynchus (shortfinned pilot whale)	42	25	Hall, Gilmartin, and Mattson, 1971; Arima and Nagakura, 1979; O'Shea et al., 1980; Tanabe et al., 1987, 1998
Grampus griseus (Risso's dolphin)	42	35	Jarman et al., 1996a; Kim et al., 1996b; Prudente et al., 1997; Hayteas and Duffield, 1997
Kogia breviceps (pygmy sperm whale)	—	2	Kemper et al., 1994; Tanabe et al., 1998
Kogia simus (dwarf sperm whale)	—	1	Tanabe et al., 1998
Lagenodelphis hosei (Fraser's dolphin)	10	1	O'Shea et al., 1980; Prudente et al., 1997; Tanabe et al., 1996, 1998
Lagenorhynchus obliquidens (Pacific whitesided dolphin)	11	—	Prudente et al., 1997; Tanabe et al., 1983, 1996
Lagenorhynchus obscurus (dusky dolphin)	1	1	Koeman et al., 1972, 1975; Tanabe et al., 1983
Lissodelphis borealis (northern right-whale dolphin)	2	2	Prudente et al., 1997; Tanabe et al., 1996
Mesoplodon ginkgodens (ginkgo-toothed whale)	—	2	Iwata et al., 1994; Tanabe et al., 1998
Mesoplodon stejnegeri (Stejneger's beaked whale)	2	1	Hayteas and Duffield, 1997; Tanabe et al., 1998; Miyazaki et al., 1987
Mesoplodon sp.	2	1	Kemper et al., 1994
Neophocoena phocaenoides (finless porpoise)	6	13	Arima and Nagakura, 1979; O'Shea et al., 1980; Kannan et al., 1989; Iwata et al., 1994, 1995; Tanabe et al., 1998
Orcinus orca (killer whale)	9	4	Ono et al., 1987; Kannan et al., 1989; Iwata et al., 1994; Jarman et al., 1996a; Tanabe et al., 1998

Species (common name)			References
Peponocephala electra (melonheaded whale)	11	—	Tanabe et al., 1983, 1996; Prudente et al., 1997
Phocoena phocoena (harbour porpoise)	77	—	Calambokidis and Barlow, 1991; Varanasi et al., 1993b; Jarman et al., 1996a; Tanabe et al., 1997; Hayteas and Duffield, 1997
Phocoenoides dalli (Dall's porpoise)	104	18	O'Shea et al., 1980; Hamanaka and Mishima, 1981; Tanabe et al., 1983, 1994, 1996, 1998; Subramanian et al., 1987; Subramaniam, Tanabe, and Tatsukawa, 1988b; Fujise et al., 1988; Kawano et al., 1988; Kannan et al., 1989; Iwata et al., 1994; Jarman et al., 1996a; Prudente et al., 1997; Hayteas and Duffield, 1997
Physeter macrocephalus (sperm whale)	9	9	Wolman and Wilson, 1970; Nagakura et al., 1974; Schafer et al., 1984; Anderson, 1991; Kemper et al., 1994
Pseudorca crassidens (false killer whale)	4	39	Baird, Langelier, and Stacey, 1989; Bergman et al., 1994; Kemper et al., 1994; Jarman et al., 1996a
Sousa chinensis (Pacific humpback dolphin)	—	13	Parsons, 1998; Tanabe et al., 1998
Stenella attenuata (pantropical spotted dolphin)	1	75	Arima and Nagakura, 1979; André et al., 1990, 1991; Kemper et al., 1994
Stenella coeruleoalba (striped dolphin)	69	247	Arima and Nagakura, 1979; O'Shea et al., 1980; Tanabe et al., 1981, 1982, 1983, 1996; Tanabe, Tanaka, and Tatsukawa, 1984; Honda et al., 1983, 1986; Honda and Tatsukawa, 1983; Itano et al., 1984a, 1984b, 1984c; Itano, Kawai, and Tatsukawa, 1985a, 1985b; Kawai et al., 1988; Loganathan et al., 1990; Prudente et al., 1997; Hayteas and Duffield, 1997
Stenella longirostris (spinner dolphin)	7	—	Prudente et al., 1997; Tanabe et al., 1996, 1998
Stenella sp.	—	3	Kemper et al., 1994
Steno bredanensis (rough-toothed dolphin)	7	6	O'Shea et al., 1980; Tanabe et al., 1998
Tursiops truncatus (bottlenose dolphin)	15	15	Arima and Nagakura, 1979; O'Shea et al., 1980; Schafer et al., 1984; Kemper et al., 1994; Tanabe et al., 1998

Note: Organotin determinations are included with metals. Scientific names are given with common names in parentheses, following the nomenclature of Rice (1998)

Table 9.4 Summaries of number of individual baleen whales, dugongs, and sea otters sampled for organochlorine and metal contamination in the Pacific Ocean, based on publications up until the end of 1998

Species	Organochlorines	Metals	Principal references
Balaenoptera acutorostrata (northern minke whale)	32	—	Schafer *et al.*, 1984; Varanasi *et al.*, 1993b; Hayteas and Duffield, 1997; Aono *et al.*, 1997
Balaenoptera borealis (sei whale)	—	9	Nagakura *et al.*, 1974
Balaenoptera edeni (Bryde's whale; Eden's whale)	2	—	Pantoja *et al.*, 1984, 1985
Balaenoptera musculus (blue whale)	1	—	Britt and Howard, 1983
Balaenoptera physalus (fin whale)	1	—	Pantoja *et al.*, 1984, 1985
Balaenoptera sp.	—	5	Kemper *et al.*, 1994
Caperea marginata (pygmy right whale)	1	1	Kemper *et al.*, 1994
Eschrichtius robustus (grey whale)	48	11	Wolman and Wilson, 1970; Schafer *et al.*, 1984; Varanasi *et al.*, 1993a, 1994
Dugong dugon (dugong)	2	47	Miyazaki *et al.*, 1979; Denton *et al.*, 1980; Denton and Breck, 1981
Enhydra lutris (sea otter)	>70	58	Shaw, 1971; Smith *et al.*, 1990; Jarman *et al.*, 1996b; Estes *et al.*, 1997; Kannan *et al.*, 1998; Nakata *et al.*, 1998

Note: Organotin determinations are included with metals. Scientific names are given with common names in parentheses, following the nomenclature of Rice (1998)

the cyclodienes aldrin, endrin, dieldrin, heptachlor, heptachlor epoxide, endrin, and various isomers of chlordane, as well as toxaphene, HCHs, HCB, and mirex (e.g. Nakata *et al.*, 1998; O'Shea *et al.*, 1980; Tanabe *et al.*, 1996). Concentrations of organochlorine pesticides and metabolites are variable, however, according to chemical compound, species, location, age, and sex. In many samples amounts reported in blubber are quite low.

Many organochlorines of industrial origin have also been reported in tissues of marine mammals of the Pacific Ocean, including most of the many individual congeners of PCBs, PCDDs, and PCDFs known from marine mammals elsewhere, as well as octachlorostyrenes, polychorinated terphenyls and phenols, and tris-(chlorophenyl) methanol (Kannan *et al.*, 1989; Walker *et al.*, 1989; Jarman *et al.*, 1996a; Becker *et al.*, 1997; Young *et al.*, 1998).

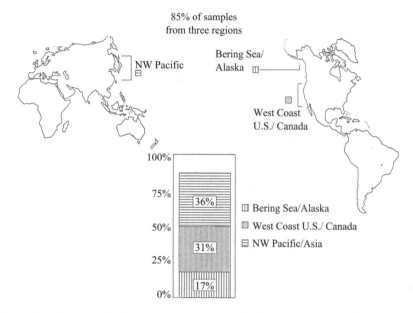

Figure 9.1 Most individual marine mammals sampled for determination of contaminant residue concentrations in the Pacific Ocean have been from three areas: the north-west Pacific near Asia, the Bering Sea and coastal Alaska, and the west coast of Canada and the USA

Elements detected in tissues of marine mammals in the Pacific Ocean have included silver (Ag), aluminium (Al), arsenic (As), bromine (Br), calcium (Ca), cadmium (Cd), chlorine (Cl), cobalt (Co), chromium (Cr), cesium (Cs), copper (Cu), iron (Fe), potassium (K), mercury (Hg), iodine (I), magnesium (Mg), manganese (Mn), sodium (Na), nickel (Ni), lead (Pb), rubidium (Rb), antimony (Sb), selenium (Se), strontium (Sr), vanadium (V), and zinc (Zn) (e.g. Martin *et al.*, 1976; Fujise *et al.*, 1988; Zeisler *et al.*, 1993; Varanasi *et al.*, 1994; Mackey *et al.*, 1996; Becker *et al.*, 1997). Organotin contamination first became known in marine mammals from analysis of liver samples from Pacific Ocean populations (Iwata *et al.*, 1994). Since then organotins have been reported from livers of marine mammals from other oceans as well, at concentrations comparable to those seen in Pacific Ocean marine mammals. Butyltins appear to be higher in livers of cetaceans than pinnipeds (Tanabe *et al.*, 1998), due at least in part to the ability of some pinnipeds to shed butyltins in their hair during the moult (Kim *et al.*, 1996c).

There has not yet been strong evidence that potentially toxic elements have caused problems in marine mammals in the Pacific Ocean or elsewhere. For marine mammals in general, including species sampled from the Pacific Ocean, the principal findings on metals have been that mercury reaches high concentrations in livers, but appears to be physiologically detoxified in reactions involving selenium; certain other organs are storage sites for other elements, such as lead in bone; and cadmium is known to increase with age in kidneys (see review in O'Shea, 1999). Cadmium toxicity, however, seems to be ameliorated by metallothionein proteins. Such findings on elements were reported for a number of Pacific Ocean populations of marine mammals, including Steller's sea-lions (*Eumetopias jubatus*) (Hamanaka, Itoo, and Mishima, 1982), harbour seals (*Phoca vitulina*) (Tohyama *et al.*, 1986), northern fur seals (*Callorhinus ursinus*) (Goldblatt and Anthony, 1983), and striped dolphins (*Stenella coeruleoalba*) (Honda and Tatsukawa, 1983).

Few marine mammals have been sampled for radionuclides anywhere in the world. In the Pacific Ocean, Calmet, Woodhead, and André (1992) reported Cs-137, K-40, and Pb-210 in 52 pantropical spotted dolphins (*Stenella attenuata*), spinner dolphins (*Stenella longirostris*), and common dolphins (*Delphinus delphis*) sampled from the eastern Pacific in the late 1970s and early 1980s. Osterberg, Pearcy, and Kiyala (1964) also measured a variety of radionuclides in a fin whale (*Balaenoptera physalus*) obtained in Oregon in the early 1960s. Radionuclide levels found were not considered harmful to the marine mammals, or indicative of biomagnification.

Temporal and spatial trends

Comparative analyses of temporal or geographic trends in contaminants in Pacific Ocean marine mammals are difficult for several reasons. First, sample sizes within species are usually too small for thorough comparisons (Tables 9.2–9.4). Second, where sample sizes seem large, they are often scattered across multiple studies, sometimes many years apart. Potential differences among these studies in analytical methodology, sample sex, age composition, and sampling locations are such that in only a few cases can even qualitative comparisons be made. Sex and age composition are very important to establish in such studies. For some organochlorines, concentrations in blubber increase in males throughout life, but not in females, who transfer them to their offspring through the lipid-rich milk beginning when the females reach reproductive age (see summaries compiled in O'Shea, 1999).

Comparisons of contaminants in marine mammals are also most valid when they are based on studies carried out in a single laboratory by the same investigators. Only a limited number of such cases can be highlighted. Among studies of baleen whales, Aono *et al.* (1997) compared organochlorine concentrations in blubber of North Pacific minke whales (*Balaenoptera acutorostrata*) with Antarctic minke whales (*Balaenoptera bonaerensis*). Minke whales from the North Pacific were higher in total DDT, PCBs, chlordanes, and HCHs, but not in HCB. β-HCH predominated in minke whales of the North Pacific and γ-HCH in Antarctic minke whales, suggesting more recent use of the pesticide lindane (γ-HCH) in the southern hemisphere. Higher organochlorine concentrations in blubber of minke whales in the North Pacific probably were due at least in part to differences in feeding strategies. North Pacific minke whales fed on fish, which Aono *et al.* (1997) found were higher in organochlorines than the krill fed upon by Antarctic minke whales. Within the North Pacific sample, whales sampled in 1994 had lower total DDT in blubber than those sampled in 1987.

At least 26 species of odontocete cetaceans have been examined for metals, organochlorines, or both (Table 9.3). However, sample sizes within species tend to be small. Wide-ranging species of odontocetes commonly sampled in other oceans, such as harbour porpoises (*Phocoena phocoena*) or bottlenose dolphins (*Tursiops truncatus*), are not well represented in collections from the Pacific Ocean, making comparisons with odontocetes sampled from other oceans not possible. Fairly large samples exist for Dall's porpoises (*Phocoenoides dalli*) and striped dolphins from the Pacific Ocean (Table 9.3), but the author found difficulties in trying to make comparisons among studies. Numbers of Dall's porpoises sampled for organochlorines, for example, are based on about nine studies at about 12 locations over at least a 20-year period. Similarly, the data on organochlorine contamination in striped dolphins are from 10 studies over 20 years at widely varying locations. Loganathan *et al.* (1990), however, compared organochlorines in blubber of adult male striped dolphins of similar age collected from the Pacific coast of Japan in 1978–1979 and again in 1986. PCBs and DDT did not change, but decreases were found in HCHs and HCB. In general changes in contamination of these marine mammals are expected to be very slow because of the diffuse sources of contamination, although major decreases in HCH use in China may have influenced decreases seen in the striped dolphins (Loganathan *et al.*, 1990).

Among Pacific Ocean pinnipeds (Table 9.2) the generalization of small numbers of scattered samples does not hold true for northern fur seals. Tanabe *et al.* (1994) and Iwata *et al.* (1998) reported organochlorine data for up to 105 fur seals of known age and sex, sampled between 1971 and 1988 in the North Pacific. Maximum concentrations of DDT and PCBs occurred around 1976 and then decreased. During the 1980s PCBs appeared to become stable, whereas total DDT continued to decline. HCHs seemed to decline more slowly. Comparisons of northern fur seals with other North Pacific marine mammals by the same laboratory showed that this species was less contaminated than Dall's porpoise, Steller's sea-lion, and the larga seal (*Phoca largha*). These pinnipeds also had lower concentrations of PCB congeners with lower numbers of chlorine atoms and lower concentrations of HCHs than Dall's porpoise, suggesting a reduced capacity to metabolize some of these compounds in the small cetacean. For other species of pinnipeds, those with larger sample sizes (Table 9.2) are spread too thinly across time, geography, and analytical laboratories to allow useful comparisons, although for California sea-lions (*Zalophus californianus*) in southern California total DDT seems to be declining (Lieberg-Clark *et al.*, 1995).

Among other groups of marine mammals, sea otters and dugongs have also been sampled for contaminants. Sea otters have not been found with excessive concentrations of organochlorines, although recent reports of concentrations of

PCBs in livers of sea otters from the remote Aleutian Islands were comparable to those from California (Estes *et al.*, 1997). California sea otters also have elevated butyltins in livers (particularly females) and butyltins tend to be higher in diseased otters, perhaps suggesting some association with immune suppression (Kannan *et al.*, 1998). Dugongs are herbivores and are not subject to exposure to organochlorines through food chain biomagnification. Concentrations of organochlorines were not high in the two individuals analysed by Miyazaki *et al.* (1979). Metal status of dugongs, however, can be more complicated because marine plants may take up different elements according to their availability in sediments and imbalances in elements may occur in different substrates (Denton *et al.*, 1980). However, harmful exposures have not been reported. No data are available to evaluate temporal or spatial trends in contamination of dugongs.

Within the Pacific Ocean, contaminants may be higher in certain areas than others. This has been shown in studies within the same laboratories on the same or similar species. For example, in one early study of organochlorines in small cetaceans sampled in the 1970s (O'Shea *et al.*, 1980), greater concentrations of certain compounds and greater numbers of compounds could be detected in small cetaceans from near-shore areas such as coastal southern California or the Seto Inland Sea, in comparison with open ocean locations. Near-shore bottlenose dolphins in coastal California had up to 2,500 ppm (wet weight) DDE in blubber. Other workers have also reported higher contamination of marine mammals in coastal areas and these have been extended to include PCDDs and PCDFs (Kannan *et al.*, 1989) and most recently butyltins. Butyltins are higher in marine mammals of coastal Pacific waters rather than the open ocean, especially around developed nations (Iwata *et al.*, 1994; Tanabe *et al.*, 1998; Tanabe, 1999).

Evidence for impact of contaminants on Pacific Ocean
marine mammals

Unlike studies of marine mammals from the Atlantic (e.g. Bergman and Olsson, 1985; Reijnders, 1986; Fossi *et al.*, 1992; Béland *et al.*, 1993; Aguilar and Borrell, 1994; Olsson, Karlsson, and Ahnland, 1994; Ross *et al.*, 1996), there have been few observations of negative phenomena in Pacific Ocean marine mammal populations that may be linked to contaminants. Watanabe *et al.* (1989) showed elevated P450 enzymes in wild cetaceans in the Pacific, but the toxicological significance of these elevated biomarkers is difficult to judge without additional data on reproduction, endocrinology, or disease. Tanabe *et al.* (1994) showed that elevated P450 enzymes in northern fur seals of the Pacific were correlated with PCB residues.

The possible roles of contaminants as immunosuppressive agents that increase susceptibility to diseases in marine mammals and their possible effects on reproduction are two areas of great concern supported by experimental evidence from feeding studies of captive harbour seals in Europe (e.g. Reijnders, 1986; DeSwart *et al.*, 1994; Ross *et al.*, 1996). Morbillivirus epizootics in cetaceans and pinnipeds in Atlantic coastal waters have both positive and negative evidence that organochlorines played a role in susceptibility to these diseases (e.g. Hall *et al.*, 1992; Aguilar and Borrell, 1994). However, there has not yet been any outstanding evidence to suggest that disease outbreaks may have occurred as a result of contaminant exposure in Pacific Ocean marine mammals. There has been speculation that elevated butyltins may be associated with various illnesses in California sea-lions (Kannan *et al.*, 1998), and elevated PCBs and DDE in serum of northern elephant seals (*Mirounga angustirostris*) from California were associated with a skin disease syndrome that shared similarities with lesions seen in other mammals suffering from PCB toxicosis (Beckmen *et al.*, 1997).

The extent of observations associating contaminants with possible reproductive disorders in marine mammals of the Pacific Ocean is very limited. In the late 1960s and early 1970s observations were made of premature births in California sea-lions associated with high DDE and possibly PCBs (DeLong, Gilmartin, and Simpson, 1973). These observations had confounding factors that made interpretation difficult (most recently reviewed by O'Shea and Brownell, 1998). These included the presence of abortion-inducing diseases and possible age-related associations. There has also been a correlation established between reduced testosterone levels in blood and DDE (but not PCB) concentrations in blubber of Dall's porpoises from the Pacific (Subramanian *et al.*, 1987).

Frontiers for research

The most important avenues for future research on marine mammals and environmental contaminants were defined by scientists participating at an expert interdisciplinary international workshop held in Keystone, Colorado in October 1998 (O'Shea, Reeves, and Kirk Long, 1999). The most important unanswered questions on this topic are global in scope and thus the critical frontiers for research in the Pacific Ocean are identical to those defined by workshop participants for marine mammals in all the oceans of the world. The areas for future research defined at that workshop are therefore excerpted later. Much of the focus of the workshop was on the persistent organochlorine contaminants. Although some countries no longer apply or produce some of these compounds, many nations continue to manufacture and use organochlorines, or allow them to be disposed of in pathways that subsequently redistribute them to the global marine ecosystem. Marine mammals at the top of complex ocean food webs then become an ultimate sink for these substances. The workshop participants were therefore more limited in their consideration of potentially toxic trace elements and other persistent substances. This is also in part because thus far there is much less evidence suggesting direct harm to marine mammals from these other contaminants.

Participants at the Keystone workshop grouped the frontiers for research into four broad conceptual topics: reproduction and endocrinology; immunotoxicity, pathology, and disease; risk assessment; and future trends. In the area of reproduction and endocrinology, the workshop recognized four major points for future research. One is to develop a better understanding of how organochlorines may influence reproduction at pre-implantation stages, as suggested by experiments on captive harbour seals in Europe (Reijnders, 1986). A second is to determine if and how organochlorines may be involved in causing premature births in pinnipeds, as suggested by observations on California sea lions (DeLong, Gilmartin, and Simpson, 1973). The third point was to design and measure indices of developmental and reproductive functions in marine mammals. These species are exposed to endocrine-disrupting chemicals at some level, but to date there is little evidence for possible effects in marine mammals. This may be because scientists have not been systematically looking for anatomical endpoints of such disruption (such as changes in size, mass, morphology, or histology of sexual organs, endocrine glands, and accessory structures). The fourth point is to carry out research in multidisciplinary approaches that include pathology and other fields and that can scale up from individual to population-level effects. Such work should take place both in the field and where necessary in captive experiments.

In the area of immunotoxicity, pathology, and disease research, six points were stressed. These were to conduct research that will allow linkages to be made between exposure to contaminants and immunosuppresive endpoints; to combine these studies with other multidisciplinary work, using otherwise well-studied populations of the same species in areas of contrasting high and low contamination; to increase efforts that try to understand the actual mechanisms by which immunotoxicity may take place in marine mammals; to take multiple approaches in studying this topic using both wild and captive individuals; to attempt to obtain results that allow scaling up from the individual to the population level; and to develop standards for archiving samples and for monitoring disease, pathology, and immune function in wild populations of marine mammals.

The working groups recognized that there were formal, structured models that can be used to conduct risk assessments to judge possible impacts of contaminants on marine mammals. Such approaches have not yet been taken with marine mammals, but should be. Risk assessment procedures will require a variety of information, including knowledge of exposure, mechanisms of threat, how the contaminant is distributed in populations, how it is mobilized and sequestered in the body, and what the targets of toxicity are. Risk assessment will also require understanding dose-response relationships, perhaps best derived through non-lethal captive experiments and experiments on meaningful surrogate species of laboratory mammals that are integrated with field studies. Risk assessment models should also incorporate margins for precaution, such that conservation of species is not unduly jeopardized.

Workshop participants noted that there are hundreds of other lipophilic (differentially absorbed and sequestered in fatty tissues) contaminants that are in or may be released to the marine environment in the future; although these may reach marine mammals, their tissues are not now routinely examined for their presence. It was also noted that there are classes of compounds in the marine environment that do not reach high levels in marine mammals but nonetheless could have toxic effects; these have not yet been investigated with respect to possible impacts on marine mammals. In addition, there are threats from increasing eutrophication of the near-shore environment, which may play a role in increasing exposure of marine mammals to some biotoxins. The inshore environment is also increasingly contaminated with monomers and polymers of unknown significance to marine mammals. In addition to many unknown marine contaminants, it is expected that distributions of PCBs and organochlorine pesticides in the world's marine mammal populations will change as these contaminants continue to enter the oceans and mass balances are shifted. Researchers will also need to strive to develop and adhere to consistent analytical protocols, including determination of and reporting of specified numbers of contaminants in marine mammal samples.

Workshop participants voiced a concern that many currently known contaminants that may affect marine mammals continue to be produced, used, and disposed of in some nations in such ways that they will continue to enter the seas in the future. The workshop stressed that much more needs to be learned about the emissions and sources of contaminants that are entering the marine environment; that specimen banking (Miyazaki, 1994) be instituted to assist in the evaluation of future threats; and that chemical monitoring be more closely linked to other, more broadly based, biological studies of marine mammals. It was also stressed that results of environmental chemistry studies of marine mammals should be made available to managers in easily usable and understandable formats; that long-term monitoring be maintained; that international experts periodically meet to evaluate the threats of contaminants to marine mammals; and that an ongoing dialogue on this topic be established and continued among international authorities.

The workshop participants were concerned about the great uncertainty that remains about specific effects of contaminants in marine mammals, to what extent such effects may occur in marine mammals in the wild, and what impact such effects are having on their population dynamics. The areas of uncertainty defined by the workshop participants as having the highest priority for resolution provide frontiers for research where science can better guide policy, management, and regulatory decisions on issues of the impacts of contaminants on marine mammals. More detailed recommendations for research in these areas are available in the workshop report (O'Shea, Reeves, and Kirk Long, 1999).

For most of the past 40 years the study of marine mammals and contaminants has documented contaminant residues in tissues of many thousands of individuals. However, there has been much less emphasis placed on designing and executing studies that allow interpretation of the significance of this contamination to marine mammal health and population dynamics. A consistent theme during the workshop was the need for multidisciplinary studies that integrate physiological, behavioural, reproductive, clinical, pathologic, and toxicological data to evaluate the relationships of immune status, health, reproduction, and survival of individuals to population- and ecosystem-level trends. Such studies should be conducted on both wild and captive populations using statistically rigorous experiments and sampling designs. Long-term research and monitoring programmes are especially valuable and need to have a stable funding base. Ongoing long-term population and life-history studies of marine mammals should strive to include components on contaminants and health.

Understanding of the subcellular mechanisms by which contaminants affect marine mammals can be achieved through *in vitro* studies using marine mammal cell lines or through experimentation. Invasive experiments can use laboratory animals as surrogates for marine mammals, although variation in response among species means that results must be treated with caution. Therefore, establishment of dose–response relationships and response thresholds may require well-designed non-lethal experimentation with marine mammals. To a considerable extent, model species that have been well studied and are readily available in captivity (e.g. California sea-lions, harbour seals, bottlenose dolphins, and belugas, *Delphinapterus leucas*) can be used to represent other related species in future field studies or research on captive marine mammals, although extrapolation must be done judiciously.

Because most marine mammals are exposed to multiple contaminants in nature, experiments on cell lines, surrogate species, or captive marine mammals should include exposures to complex mixtures in addition to single chemicals. (However, focusing on single chemicals will also remain important because of the more straightforward potential for management action where such chemicals are still in use.) Biochemical and physiological indices of exposure or effects ("biomarkers") can be useful monitoring tools. More biomarkers need to be developed and validated for marine mammals. Improvements in understanding blubber and lipid physiology in relation to contaminant storage and mobilization are also critical topics. Protocols should be developed that allow assessments of morphological or other endpoints of endocrine disruption as components of ongoing stranding programmes as well as integral parts of studies of living marine mammals.

Workshop participants emphasized the potential for major problems in the future with well-known contaminants, substances not yet identified by current analyses, and many "new" contaminants that are being developed or are already in production. Thus model contaminants should be selected for intensive study based on levels of production, potential for bioaccumulation, toxicity, and suspected or known adverse effects in marine mammals. Formal, quantitative risk assessment models and procedures should be used to estimate contaminant impacts on marine mammal populations. Universal protocols should be followed for sample collection, storage, laboratory analysis and data reporting, and for maximizing the number of contaminants identified in marine mammal samples. Enhanced international cooperation and improvements to systems for compiling, interpreting, and disseminating data on persistent contaminants in marine ecosystems will be of great importance to the future. Monitoring contaminants in marine mammals on a global scale will be integral to such improvements.

ACKNOWLEDGEMENTS

Organizers and sponsors of the international conference "Man and the Oceans" invited the author to participate in the conference and to develop this chapter. Professor Nobuyuki Miyazaki kindly extended time to complete the written version and Professor Shinsuke Tanabe encouraged an expanded coverage of topics. Professor Ricardo Bastida provided useful references. The author thanks the participants at the Keystone workshop on marine mammals and persistent ocean contaminants for defining the key frontiers for research on this topic and Randy Reeves and Alison Kirk Long for helping synopsize these points in a form that allowed them to be excerpted in this chapter. The US Marine Mammal Commission, National Marine Fisheries Service, Environmental Protection Agency, National Fish and Wildlife Foundation, and US Geological Survey sponsored the latter workshop.

REFERENCES

Aguilar, A. and A. Borrell. 1994. Abnormally high polychlorinated biphenyl levels in striped dolphins (*Stenella coeruleoalba*) affected by the 1990–1992 Mediterranean epizootic. *Science of the Total Environment*, 154: 237–247.

Anas, R.E. 1971. Organochlorine pesticides in northern fur seals, California sea lions, and birds, 1968-69. In: *Fur Seal Investigations, 1969*, pp. 32–36. National Marine Fisheries Service, US Department of Commerce, Special Scientific Report Fisheries No. 628.

Anas, R.E. 1973. Mercury in fur seals. In: *Proceedings of the Workshop on Mercury in the Western Environment*, Donald R. Buhler (ed.), pp. 91–96. Oregon State University Continuing Education Publications, Corvallis, Oregon.

Anas, R.E. 1974a. Heavy metals in the northern fur seal, *Callorhinus ursinus*, and harbour seals, *Phoca vitulina richardi. Fishery Bulletin*, 72: 133–137.

Anas, R.E. 1974b. DDT plus PCBs in blubber of harbour seals. *Pesticides Monitoring Journal*, 8: 12–14.

Anas, R.E. and A.J. Wilson. 1970. Organochlorine pesticides in fur seals. *Pesticides Monitoring Journal*, 3: 198–200.

Anas, R.E. and D.D. Worlund. 1975. Comparison between two methods of subsampling blubber of northern fur seals for total DDT plus PCBs. *Pesticides Monitoring Journal* 8: 261–262.

Anderson, G.R.V. 1991. Australia. Progress Report on Cetacean Research, May 1989 to May 1990. Reports of the International Whaling Commission, 41: 223–229.

André, J.M., J.C. Amiard, C. Amiard-Triquet, A. Boudou, and F. Ribeyre. 1990. Cadmium contamination of tissues and organs of delphinids Species (*Stenella attenuata*) – Influence of biological and ecological factors. *Ecotoxicology and Environmental Safety*, 20: 290–306.

André, J.M., A. Boudou, and F. Ribeyre. 1991. Mercury accumulation in *Delphinidae. Water, Air, and Soil Pollution*, 56: 187–201.

Aono, S., S. Tanabe, Y. Fujise, H. Kato, and R. Tatsukawa. 1997. Persistent organochlorines in minke whale (*Balaenoptera acutorostrata*) and their prey species from the Antarctic and the North Pacific. *Environmental Pollution*, 98: 81–89.

Arima, S. and K. Nagakura. 1979. Mercury and selenium content of *Odontoceti*. *Bulletin of the Japanese Society of Scientific Fisheries*, 45: 623–626.

Bacon, C.E., W.M. Jarman, and D.P. Costa. 1992. Organochlorine and polychlorinated biphenyl levels in pinniped milk from the Arctic, the Antarctic, California and Australia. *Chemosphere*, 24: 779–791.

Baird, R.W., K.M. Langelier, and P.J. Stacey. 1989. First records of false killer whales, *Pseudorca crassidens*, in Canada. *Canadian Field-Naturalist*, 103: 368–371.

Becker, P.R., E.A. Mackey, R. Demiralp, R. Suydam, G. Early, B.J. Koster, and S.A. Wise. 1995. Relationship of silver with selenium and mercury in the liver of two species of toothed whales (Odontocetes). *Marine Pollution Bulletin*, 30: 262–271.

Becker, P.R., E.A. Mackey, R. Demiralp, M.M. Schantz, B.J. Koster, and S.A. Wise, 1997. Concentrations of chlorinated hydrocarbons and trace elements in marine mammal tissues archived in the U.S. national biomonitoring specimen bank. *Chemosphere*, 34: 2067–2098.

Beckmen, K.B., L.J. Lowenstine, J. Newman, J. Hill, K. Hanni, and J. Gerber. 1997. Clinical and pathological characterization of northern elephant seal skin disease. *Journal of Wildlife Diseases*, 33: 438–439.

Béland, P., S. DeGuise, C. Girard, A. Lagacé, D. Martineau, R. Michaud, D.C.G. Muir, R.J. Norstrom, É. Pelletier, S. Ray, and L.R. Shugart. 1993. Toxic compounds and health and reproductive effects in St. Lawrence Beluga Whales. *Journal of Great Lakes Research*, 19: 766–775.

Bergman, Å., R.J. Norstrom, K. Haraguchi, H. Kuroki, and P. Béland. 1994. PCB and DDE methyl sulfones in mammals from Canada and Sweden. *Environmental Toxicology and Chemistry*, 13: 121–128.

Bergman, A. and M. Olsson. 1985. Pathology of Baltic grey seal and ringed seal females with special reference to adrenocortical hyperplasia: Is environmental pollution the cause of a widely distributed disease syndrome? *Finnish Game Research*, 44: 47–62.

Braham, H.W. 1973. Lead in the California sea-lion (*Zalophus californianus*). *Environmental Pollution*, 5: 253–258.

Britt, J.O. and E.B. Howard. 1983. Tissue residues of selected environmental contaminants in marine mammals. In: *Pathobiology of Marine Mammal Diseases*, Edwin B. Howard (ed.), II: pp. 79–94. CRC Press, Inc., Boca Raton, Florida.

Buhler, D.R., R.R. Claeys, and B.R. Mate. 1975. Heavy metal and chlorinated hydrocarbon residues in California sea-lions (*Zalophus californianus californianus*). *Journal of the Fisheries Research Board of Canada*, 32: 2391–2397.

Calambokidis, J. and J. Barlow. 1991. Chlorinated hydrocarbon concentrations and their use for describing populations discreteness in harbor porpoises from Washington, Oregon, and California. *Marine Mammal Strandings in the United States: Proceedings of the Second Marine Mammal Stranding Workshop*, J.E. Reynolds and D.K. Odell (eds) pp. 101–110. U.S. Department of Commerce, National Oceanic and Atmospheric Administration Technical Report NMFS 98. Washington, DC.

Calmet, D., D. Woodhead, and J.M. André. 1992. ^{210}Pb, ^{137}Cs and ^{40}K in three species of porpoises caught in the eastern tropical pacific ocean. *Journal of Environmental Radioactivity*, 15: 153–169.

DeLong, R.L., W.G. Gilmartin, and J.G. Simpson. 1973. Premature births in California sea lions: association with high organochlorine pollutant residue levels. *Science*, 181: 1168–1170.

Denton, G.R.W. and W.G. Breck. 1981. Mercury in tropical marine organisms from North Queensland. *Marine Pollution Bulletin*, 12: 116–121.

Denton, G.R.W., H. Marsh, G.E. Heisohn, and C. Burdon-Jones. 1980. The unusual metal status of the dugong (*Dugong dugon*). *Marine Biology*, 57: 201–219.

DeSwart, R.L., P.S. Ross, L.J. Vedder, H.H. Timmerman, S. Heisterkamp, H. Van Loveren, J.G. Vos, P.J.H. Reijnders, and A.D.M.E. Osterhaus. 1994. Impairment in immune function in harbour seals (*Phoca vitulina*) feeding on fish from polluted waters. *AMBIO*, 23: 155–159.

Estes, J. A., C.E. Bacon, W.M. Jarman, R.J. Norstrom, R.G. Anthony, and A.K. Miles. 1997. Organochlorines in sea otters and bald eagles from the Aleutian archipelago. *Marine Pollution Bulletin*, 34: 486–490.

Fossi, M.C. and L. Marsili. 1997. The use of non-destructive biomarkers in the study of marine mammals. *Biomarkers*, 2: 205–216.

Fossi, M.C., L. Marsili, C. Leonzio, G. Notarbartolo di Sciara, M. Zanardelli, and S. Focardi. 1992. The use of non-destructive biomarker in Mediterranean cetaceans: Preliminary data on MFO activity in skin biopsy. *Marine Pollution Bulletin*, 24: 459–461.

Fujise, Y., K.Honda, R. Tatsukawa, and S. Mishima. 1988. Tissue distribution of heavy metals in Dall's porpoise in the northwestern Pacific. *Marine Pollution Bulletin*, 19: 226–230.

Geraci, J.R. and D.J. St. Aubin (eds). 1990. *Sea Mammals and Oil: Confronting the Risks*. Academic Press, San Diego, CA.

Goldblatt, C.J. and R.G. Anthony. 1983. Heavy metals in northern fur seals (*Callorhinus ursinus*) from the Pribilof Islands, Alaska. *Journal of Environmental Quality*, 12: 478–482.

Hall, A.J., R.J. Law, D.E. Wells, J. Harwood, H.M. Ross, S. Kennedy, C.R. Allchin, L.A. Campbell, and P.P. Pomeroy. 1992. Organochlorine levels in common seals (*Phoca vitulina*) which were victims and survivors of the 1988 phocine distemper epizootic. *The Science of the Total Environment*, 115: 145–162.

Hall, J.D., W.G. Gilmartin, and J.L. Mattsson. 1971. Investigation of a Pacific pilot whale stranding on San Clemente island. *Journal of Wildlife Diseases*, 7: 324–327.

Hamanaka, T. and S. Mishima. 1981. Cadmium and zinc concentrations in marine organisms in the northern North Pacific Ocean. In: *Research Institute of North Pacific Fisheries*, pp. 191–200. Hokkaido University Special Volume.

Hamanaka, T., H. Kato, and T. Tsujita. 1977. Cadmium and zinc in ribbon seal, *Histriophoca fasciata*, in the Okhotsk Sea. In: *Research Institute of North Pacific Fisheries*, pp. 547–561. Hokkaido University, Special Volume.

Hamanaka, T., T. Itoo, and S. Mishima. 1982. Age-related change and distribution of cadmium and zinc concentrations in the Steller sea-lion (*Eumetopias jubata*) from the coast of Hokkaido, Japan. *Marine Pollution Bulletin*, 13: 57–61.

Hayteas, D.L. and D.A. Duffield. 1997. The determination by HPLC of PCB and *p,p*-DDE residues in marine mammals stranded on the Oregon coast, 1991–1995. *Marine Pollution Bulletin*, 34: 844–848.

Himeno, S., C. Watanabe, T. Hongo, T. Suzuki, A. Naganuma, and N. Imura. 1989. Body size and organ accumulation of mercury and selenium in young harbour seals (*Phoca vitulina*). *Bulletin of Environmental Contamination and Toxicology*, 42: 503–509.

Honda, K. and R. Tatsukawa. 1983. Distribution of cadmium and zinc in tissues and organs, and their age-related changes in striped dolphins, *Stenella coeruleoalba*. *Archives of Environmental Contamination and Toxicology*, 12: 543–550.

Honda, K., R. Tatsukawa, K. Itano, N. Miyazaki, and T. Fujiyama. 1983. Heavy metal concentrations in muscle, liver and kidney tissue of striped dolphin, *Stenella coeruleoalba*, and their variations with body length, weight, age and sex. *Agricultural and Biological Chemistry*, 47: 1219–1228.

Honda, K., Y. Fujise, R. Tatsukawa, K. Itano, and N. Miyazaki. 1986. Age-related accumulation of heavy metals in bone of the striped dolphin, *Stenella coeruleoalba*. *Marine Environmental Research*, 20: 143–160.

Itano, K., S. Kawai, N. Miyazaki, R. Tatsukawa, and T. Fujiyama. 1984a. Mercury and selenium levels in striped dolphins caught off the Pacific Coast of Japan. *Agricultural and Biological Chemistry*, 48: 1109–1116.

Itano, K., S. Kawai, N. Miyazaki, R. Tatsukawa, and T. Fujiyama. 1984b. Body burdens and distribution of mercury and selenium in striped dolphins. *Agricultural and Biological Chemistry* 48: 1117–1121.

Itano, K., S. Kawai, N. Miyazaki, R. Tatsukawa, and T. Fujiyama. 1984c. Mercury and selenium levels at the fetal and suckling stages of striped dolphin, *Stenella coeruleoalba*. *Agricultural and Biological Chemistry*, 48: 1691–1698.

Itano, K., S. Kawai, and R. Tatsukawa. 1985a. Properties of mercury and selenium in salt-insoluble fraction of muscles in striped dolphin. *Bulletin of the Japanese Society of Scientific Fisheries*, 51: 1129–1131.

Itano, K., S. Kawai, and R. Tatsukawa. 1985b. Distribution of mercury and selenium in muscle of striped dolphins. *Agricultural and Biological Chemistry*, 49: 515–517.

Iwata, Hisato, Shinsuke Tanabe, Nobuyuki Miyazaki, and Ryo Tatsukawa. 1994. Detection of butyltin compound residues in the blubber of marine mammals. *Marine Pollution Bulletin*, 28: 607–612.

Iwata, H., S. Tanabe, T. Mizuno, and R. Tatsukawa. 1995. High accumulation of toxic butyltins in marine mammals from Japanese coastal waters. *Environmental Science and Technology*, 29: 2959–2962.

Iwata, H., S. Tanabe, T. Mizuno, and R. Tatsukawa. 1997. Bioaccumulation of butyltin compounds in marine mammals: The specific tissue distribution and composition. *Applied Organometallic Chemistry*, 11: 257–264.

Iwata, H., S. Tanabe, T. Iida, N. Baba, J.P. Ludwig, and R. Tatsukawa. 1998. Enantioselective accumulation of hexachlorocyclohexane in northern fur seals and double-crested cormorants: Effects of biological and ecological factors in the higher trophic levels. *Environmental Science and Technology*, 32: 2244–2249.

Jarman, W.W., R.J. Norstrom, D.C.G. Muir, B. Rosenberg, M. Simon, and R.W. Baird. 1996a. Levels of organochlorine compounds, including PCDDS and PCDFS, in the blubber of cetaceans from the west coast of North America. *Marine Pollution Bulletin*, 32: 426–436.

Jarman, W.M., C.E. Bacon, J.A. Estes, M. Simon, and R.J. Norstrom. 1996b. Organochlorine contaminants in sea otters: The sea otter as a bioindicator. *Endangered Species Update*, 13: 20–22.

Kannan, K., K.S. Guruge, N.J. Thomas, S. Tanabe, and J.P. Giesy. 1998. Butyltin residues in southern sea otters (*Enhydra lutris nereis*) found dead along California coastal waters. *Environmental Science and Technology*, 32: 1169–1175.

Kannan, N., S. Tanabe, M. Ono, and R. Tatsukawa. 1989. Critical evaluation of polychlorinated biphenyl toxicity in terrestrial and marine mammals: Increasing impact of non-ortho and mono-ortho coplanar polychlorinated biphenyls from land to ocean. *Archives of Environmental Contamination and Toxicology*, 18: 850–857.

Kawai, S., M. Fukushima, N. Miyazaki, and R. Tatsukawa. 1988. Relationship between lipid composition and organochlorine levels in the tissues of striped dolphin. *Marine Pollution Bulletin*, 19: 129–133.

Kawano, Masahide, Tsuyoshi Inoue, Toyohito Wada, Hideo Hidaka, and Ryo Tatsukawa. 1988. Bioconcentration and residue patterns of chlordane compounds in marine animals: Invertebrates, fish, mammals, and seabirds. *Environmental Science and Technology*, 22: 792–797.

Kemper, Catherine, Philip Gibbs, David Obendorf, Steven Marvanek, and Cor Lenghaus. 1994. A review of heavy metal and organochlorine levels in marine mammals in Australia. *The Science of the Total Environment*, No. 154, pp. 129–139.

Kim, Gi B., Shinsuke Tanabe, Ryo Tatsukawa, Thomas R. Loughlin, and Kenji Shimazaki. 1996a. Characteristics of butyltin accumulation and its biomagnification in Steller sea- lion (*Eumetopias jubatus*). *Environmental Toxicology and Chemistry*, 15: 2043–2048.

Kim, G.B., S. Tanabe, R. Iwakiri, R. Tatsukawa, M. Amano, N. Miyazaki, and H. Tanaka. 1996b. Accumulation of butyltin compounds in Risso's dolphin (*Grampus griseus*) from the Pacific coast of Japan: Comparison with organochlorine residue pattern. *Environmental Science and Technology*, 30: 2620–2625.

Kim, Gi B., Jong S. Lee, Shinsuke Tanabe, Hisato Iwata, Ryo Tatsukawa, and Kenji Shimazaki. 1996c. Specific accumulation and distribution of butyltin compounds in various organs and tissues of the Steller sea-lion (*Eumetopias jubatus*): Comparison with organochlorine accumulation pattern. *Marine Pollution Bulletin*, 32: 558–563.

Kim, Ke C., Richard C. Chu, and George P. Barron. 1974. Mercury in tissues and lice of northern fur seals. *Bulletin of Environmental Contamination and Toxicology*, 11: 281–284.

Koeman, J.H., W.H.M. Peters, C.J. Smit, P.S. Tjioe and J.J.M. Goeij. 1972. Persistent chemicals in marine mammals. *TNO Nieuws*, 27: 570–578.

Koeman, J.H., W.S.M. van de Ven, J.J.M. de Goeij, P.S. Tjioe, and J.L. van Haaften. 1975. Mercury and selenium in marine mammals and birds. *Science of the Total Environment*, 3: 279–287.

Krahn, Margaret M., Paul R. Becker, Karen L. Tilbury, and John E. Stein. 1997. Organochlorine contaminants in blubber of four seal species: Integrating biomonitoring and specimen banking. *Chemosphere*, 34: 2109–2121.

Kurtz, David A. and Ke C. Kim. 1976. Chlorinated hydrocarbon and PCB residues in tissues and lice of northern fur seals, 1972. *Pesticides Monitoring Journal*, 10: 79–83.

LeBouef, Burney J. and Michael L. Bonnell. 1971. DDT in California sea-lions. *Nature* 234: 108–109.

Lee, J.S., Shinsuke Tanabe, H. Umino, R. Tatsukawa, T.R. Loughlin, and D.C. Calkins. 1996. Persistent organochlorines in steller sea-lion (*Eumetopias jubatus*) from the bulk of Alaska and the Bering Sea, 1976–1981. *Marine Pollution Bulletin*, 32: 535–544.

Lee, S.S., B.R. Mate, K.T. von der Trenck, R.A. Rimerman, and D.R. Buhler. 1977. Metallothionein and the subcellular localization of mercury and cadmium in the California sea lion. *Comparative Biochemistry and Physiology*, 57C: 45–53.

Lieberg-Clark, Patricia, Corinne E. Bacon, Stephanie A. Burns, Walter M. Jarman, and Burney J. Le Boeuf. 1995. DDT in California sea-lions: A follow-up study after twenty years. *Marine Pollution Bulletin*, 30: 744–745.

Loganathan, B.G., S. Tanabe, H. Tanaka, S. Watanabe, Nobuyuki Miyazaki, M. Amano, and Ryo Tatsukawa. 1990. Comparison of organochlorine residue levels in the striped dolphin from western North Pacific, 1978-79 and 1986. *Marine Pollution Bulletin*, 21: 435–439.

Loughlin, Thomas R. (ed.). 1994. *Marine Mammals and the Exxon Valdez*. Academic Press, San Diego.

Mackey, E.A., P.R. Becker, R. Demiralp, R.R. Greenberg, B.J. Koster, and S.A. Wise. 1996. Bioaccumulation of vanadium and other trace metals in livers of Alaskan cetaceans and pinnipeds. *Archives of Environmental Contamination and Toxicology*, 30: 503–512.

Martin, J.H., P.D. Elliott, V.C. Anderlini, D. Girvin, S.A. Jacobs, R.W. Risebrough, R.L. Delong and W.G. Gilmartin. 1976. Mercury–selenium–bromine imbalance in premature parturient California sea-lions. *Marine Biology*, 35: 91–104.

Mead, James G. and Robert L. Brownell, Jr. 1993. Order Cetacea. In: *Mammal Species of the World: A Taxonomic and Geographic Reference*, Second Edition. Don E. Wilson and DeeAnn M. Reeder (eds), pp. 349–364. Smithsonian Institution Press, Washington, DC.

Miles, A. Keith, Donald G. Calkins, and Nancy C. Coon. 1992. Toxic elements and organochlorines in harbour seals (*Phoca vitulina richardsi*), Kodiak, Alaska, USA. *Bulletin of Environmental Contamination and Toxicology*, 48: 727–732.

Miyazaki, Nobuyuki. 1994. Contaminant monitoring studies using marine mammals and the need for establishment of an international environmental specimen bank. *Science of the Total Environment*, 154: 249–256.

Miyazaki, Nobuyuki, K. Itano, M. Fukushima, S.-I. Kawai, and K. Honda. 1979. Metals and organochlorine compounds in the muscle of dugong from Sulawesi Island. *Scientific Reports of the Whales Research Institute* (Japan), 31: 125–128.

Miyazaki, Nobuyuki, Izumi Nakamura, Shinsuke Tanabe, and Ryo Tatsukawa. 1987. A stranding of *Mesoplodon stejnegeri* in the Maizuru Bay, Sea of Japan. *Scientific Reports of the Whales Research Institute*. Tokyo, 38: 91–105.

Mössner, Stephanie and Karlheinz Ballschmiter. 1997. Marine mammals as global pollution indicators for organochlorines. *Chemosphere*, 34: 1285–1296.

Mössner, Stephanie, T.R. Spraker, P.R. Becker, and Karlheinz Ballschmiter. 1992. Ratios of enantiomers of alpha-HCH and determination of alpha-, beta-, and gamma-HCH isomers in brain and other tissues of neonatal northern fur seals (*Callorhinus ursinus*). *Chemosphere*, 24: 1171–1180.

Nagakura, Katsuo, Satoshi Arima, Michihiko Kurihara, Takeshi Koga, and Takeo Fujita. 1974. Mercury content of whales. *Bulletin of Tokai Registry of Fisheries Research Laboratory*, 78: 41–46.

Nakata, H., K. Kannan, L. Jing, N. Thomas, S. Tanabe, and J.P. Giesy. 1998. Accumulation pattern of organochlorine pesticides and polychlorinated biphenyls in southern sea otters (*Enhydra lutris nereis*) found stranded along coastal California, USA. *Environmental Pollution*, 103: 45–53.

Newman, J.W., J. Vedder, W.M. Jarman, and R.R. Chang. 1994. A method for the determination of environmental contaminants in living marine mammals using microscale samples of blubber and blood. *Chemosphere*, 28: 1795–1805.

Noda, K., H. Ichihashi, T.R. Loughlin, N. Baba, M. Kiyota, and R. Tatsukawa. 1995. Distribution of heavy metals in muscle, liver and kidney of northern fur seal (*Callorhinus ursinus*) caught off Sanriku, Japan and From the Pribilof Islands, Alaska. *Environmental Pollution*, 90: 51–59.

Olsson, Matts, Börje Karlsson, and Eva Ahnland. 1994. Diseases and environmental contaminants in seals from the Baltic and the Swedish west coast. *The Science of the Total Environment*, No. 154, pp. 217–227.

Ono, Mitsuhiro, Narayanan Kannan, Tadaaki Wakimoto, and Ryo Tatsukawa. 1987. Dibenzofurans a greater global pollutant than dioxins? Evidence from analyses of open ocean killer whale. *Marine Pollution Bulletin*, 18: 640–643.

O'Shea, Thomas J. 1999. Environmental contaminants and marine mammals. In: *Biology of Marine Mammals*, John E. Reynolds, III and Sentiel A. Rommel (eds), pp. 485–564. Smithsonian Institution Press, Washington, DC.

O'Shea, Thomas J. and Robert L. Brownell. 1998. California sea-lion (*Zalophus californianus*) populations and DDT contamination. *Marine Pollution Bulletin*, 36: 159–164.

O'Shea, Thomas J. and Shinsuke Tanabe. 2003. Persistent ocean contaminants and marine mammals: a retrospective overview. In: *Toxicology of Marine Mammals*, J.G. Vos, G.D. Bossart, M. Fournier, and T.J. O'Shea (eds). pp. 99–134. Taylor & Francis Publishers, London.

O'Shea, Thomas J., Randall R. Reeves, and Alison Kirk Long (eds). 1999. *Marine Mammals and Persistent Ocean Contaminants: Proceedings of the Marine Mammal Commission Workshop, Keystone, Colorado, 12–15 October 1998.* US Marine Mammal Commission, Bethesda, Maryland, USA.

O'Shea, Thomas J., Robert L. Brownell, Jr., Donald R. Clark Jr., William A. Walker, Martha L. Gay, and Thair G. Lamont. 1980. Organochlorine pollutants in small cetaceans from the Pacific and South Atlantic oceans, November 1968–June 1976. *Pesticides Monitoring Journal*, 14: 35–46.

Osterberg, Charles, William Pearcy, and Norman Kujala. 1964. Gamma emitters in a fin whale. *Nature*, 204: 1006–1007.

Pantoja, Silvio, Luis Pastene, Jose Becerra, Mario Silva, and Victor A. Gallardo. 1984. DDTs in balaenopterids (Cetacea) from the Chilean coast. *Marine Pollution Bulletin*, 15: 451.

Pantoja, S., L. Pastene, J. Becerra, M. Silva, and V.A. Gallardo. 1985. Lindane, aldrin and dieldrin in some Chilean cetacea. *Marine Pollution Bulletin*, 16: 255.

Parsons, E.C.M. 1998. Trace metal pollution in hong kong: implications for the health of Hong Kong's Indo-Pacific hump-backed dolphins (*Sousa chinensis*). *Science of the Total Environment*, No. 214, pp. 175–184.

Prudente, Maricar, Shinsuke Tanabe, Mafumi Watanabe, Annamalai Subramanian, Nobuyuki Miyazaki, Pauline Suarez, and Ryo Tatsukawa. 1997. Organochlorine contamination in some odontoceti species from the North Pacific and Indian Ocean. *Marine Environmental Research*, 44: 415–427.

Reijnders, Peter J.H. 1986. Reproductive failure in common seals feeding on fish from polluted coastal waters. *Nature*, 324: 456–457.

Rice, Dale W. 1998. *Marine Mammals of the World: Systematics and Distribution.* Special Publication 4, Society for Marine Mammalogy, Lawrence, Kansas, USA 231pp.

Richard, C.A. and E.J. Skoch. 1986. Comparison of heavy metal concentrations between specific tissue sites in the northern fur seal. *Proceedings of the Annual*

International Association for Aquatic Animal Medicine Conference and Workshop, 17: 94–103.

Ross, P.S., R.L. De Swart, H.H. Timmerman, P.J.H. Reijnders, J.G. Vos, H. Van Loveren, and A.D.M.E. Osterhaus. 1996. Suppression of natural killer cell activity in harbour seals (*Phoca vitulina*) fed baltic sea herring. *Aquatic Toxicology*, 34: 71–84.

Schafer, Henry A., Richard W. Gossett, Charles F. Ward, and Alvin M. Westcott. 1984. Chlorinated hydrocarbons in marine mammals. In: *Southern California Coastal Water Research Project Biennial Report, 1983–1984*, W. Bascom (ed.), pp. 109–114. Southern California Coastal Water Research Project, Long Beach, California.

Schantz, Michelle M., Barbara J. Koster, Stephen A. Wise, and Paul R. Becker. 1993. Determination of PCBs and chlorinated hydrocarbons in marine mammal tissues. *The Science of the Total Environment*, No. 139/140, pp. 323–345.

Sepúlveda, María. S., Hugo Ochoa-Acuña, and Stephen F. Sundlof. 1997. Heavy metal concentrations in Juan Fernandez fur seals (*Arctocephalus philippii*). *Marine Pollution Bulletin*, 34: 663–665.

Serat, William F., Min K. Lee, Albert J. Van Loon, Donald C. Mengle, James Ferguson, John M. Burks, and Thomas R. Bender. 1977. DDT and DDE in the blood and diet of eskimo children from Hooper Bay, Alaska. *Pesticides Monitoring Journal*, 11: 1–4.

Shaw, Stanton B. 1971. Chlorinated hydrocarbon pesticides in California sea otters and harbour seals. *California Fish and Game*, 57: 290–294.

Smith, Donald R., Sidney Niemeyer, James A. Estes, and A. Russell Flegal. 1990. Stable lead isotopes evidence anthropogenic contamination in Alaskan sea otters. *Environmental Science and Technology*, 24: 1517–1521.

Subramanian, A.N., S. Tanabe, R. Tatsukawa, S. Saito, and N. Miyazaki. 1987. Reduction in the testosterone levels by PCBs and DDE in Dall's porpoises of north-western North Pacific. *Marine Pollution Bulletin*, 18: 643–646.

Subramanian, Annamalai, Shinsuke Tanabe, and Ryo Tatsukawa. 1988a. Estimating some biological parameters of Baird's beaked whales using PCBs and DDE as tracers. *Marine Pollution Bulletin*, 19: 284–287.

Subramanian, A.N., S. Tanabe, and R. Tatsukawa. 1988b. Use of organochlorines as chemical tracers in determining some reproductive parameters in Dalli-type Dall's porpoise *Phocoenoides dalli*. *Marine Environmental Research*, 25: 161–174.

Sydeman, William J. and Walter M. Jarman. 1998. Trace metals in seabirds, Steller sea- lion, and forage fish and zooplankton from central California. *Marine Pollution Bulletin*, 36: 828–832.

Takei, G.H. and G.H. Leong. 1981. Macro-analytical methods used to analyze tissues of the Hawaiian monk seal, *Monachus schauinslandi*, for organochlorine pesticides, polychlorobiphenyls, and pentachlorophenol. *Bulletin of Environmental Contamination and Toxicology*, 27: 489–498.

Tanabe, S. 1999. Butyltin contamination in marine mammals– A review. *Marine Pollution Bulletin*, 39: 62–72.

Tanabe, S., R. Tatsukawa, H. Tanaka, K. Maruyama, N. Miyazaki, and T. Fujiyama. 1981. Distribution and total burdens of chlorinated hydrocarbons in bodies of striped dolphins (*Stenella coeruleoalba*). *Agricultural and Biological Chemistry*, 45: 2569–2578.

Tanabe, S., R. Tatsukawa, K. Maruyama, and N. Miyazaki. 1982. Transplacental transfer of PCBs and chlorinated hydrocarbon pesticides from the pregnant striped dolphin

(*Stenella coeruleoalba*) to her fetus. *Agricultural and Biological Chemistry*, 46: 1249–1254.

Tanabe, S., T. Mori, R. Tatsukawa, and N. Miyazaki. 1983. Global pollution of marine mammals by PCBs, DDTs and HCHs (BHCs). *Chemosphere*, 12: 1269–1275.

Tanabe, S., H. Tanaka, and R. Tatsukawa. 1984. Polychlorobiphenyls, DDT, and hexachlorocyclohexane isomers in the Western North Pacific ecosystem. *Archives of Environmental Contamination and Toxicology*, 13: 731–738.

Tanabe, S., B.G. Loganathan, A.N. Subramanian, and R. Tatsukawa. 1987. Organochlorine residues in short-finned pilot whale: possible use as tracers of biological parameters. *Marine Pollution Bulletin*, 18: 561–563.

Tanabe, S., J.-K. Sung, D.-Y. Choi, N. Baba, M. Kiyota, K. Yoshida, and R. Tatsukawa. 1994. Persistent organochlorine residues in northern fur seal from the Pacific coast of Japan since 1971. *Environmental Pollution*, 85: 305–314.

Tanabe, S., P. Kumaran, H. Iwata, R. Tatsukawa, and N. Miyazaki. 1996. Enantiomeric ratios of -hexachlorocyclohexane in blubber of small cetaceans. *Marine Pollution Bulletin*, 32: 27–31.

Tanabe, S., B. Madhusree, A.A. Özturk, R. Tatsukawa, N. Miyazaki, E. Özdamar, O. Aral, O. Sanmsun, and B. Öztürk. 1997. Persistent organochlorine residues in harbour porpoise (*Phocoena phocoena*) from the Black Sea. *Marine Pollution Bulletin*, 34: 338–347.

Tanabe, S., M. Prudente, T. Mizuno, J. Hasegawa, H. Iwata, and N. Miyazaki. 1998. Butyltin contamination in marine mammals from north Pacific and Asian coastal waters. *Environmental Science & Technology*, 32: 193–198.

Taylor, D.L., S. Schliebe, and H. Metsker. 1989. Contaminants in blubber, liver and kidney tissue of pacific walruses. *Marine Pollution Bulletin*, 20: 465–468.

Theobald, J. 1973. DDT levels in the sea lion. *Journal of Zoo Animal Medicine*, 4: 23.

Tohyama, C., S. Himeno, C. Watanabe, T. Suzuki, and M. Morita. 1986. The relationship of the increased level of metallothionein with heavy metal levels in the tissue of the harbor seal (*Phoca vitulina*). *Ecotoxicology and Environmental Safety*, 12: 85–94.

Varanasi, U., J.E. Stein, W.L. Reichert, K.L. Tilbury, M.M. Krahn, and S.-L. Chan. 1992. Chlorinated and aromatic hydrocarbons in bottom sediments, fish and marine mammals in U.S. coastal waters: laboratory and field studies of metabolism and accumulation. In: *Persistent Pollutants in Marine Ecosystems*, C. H. Walker and D.R. Livingstone (eds), pp. 83–115. Pergamon Press, New York, NY.

Varanasi, U, J.E. Stein, K.L. Tilbury, J.P. Meador, C.A. Sloan, D.W. Brown, J. Calambodikis, and S.-L. Chan. 1993a. *Chemical Contaminants in Gray Whales (Eschrichtius robustus) Stranded in Alaska, Washington, and California, U.S.A.* US Department of Commerce, National Oceanic and Atmospheric Administration, National Marine Fisheries Service. NOAA Technical Memorandum NMFS-NWFSC-11. 115pp.

Varanasi, U., J.E. Stein, K.L. Tilbury, D.W. Brown, J.P. Meador, M.M. Krahn, and S.-L. Chan. 1993b. Contaminant monitoring for NMFS marine mammal health and stranding response program. In: *Coastal Zone 93 Proceedings, The Eighth Symposium on Coastal and Ocean Management*, pp. 1–15. New Orleans, LA.

Varanasi, U., J.E. Stein, K.L. Tilbury, J.P. Meador, C.A. Wigren, R.C. Clark, and S.-L. Chan. 1994. Chemical contaminants in gray whales (*Eschrichtius robustus*) stranded along the west coast of North America. *The Science of the Total Environment*, 145: 29–53.

Walker, W., R.W. Risebrough, W.M. Jarman, B.W. de Lappe, J.A. Tefft, and R.L. DeLong. 1989. Identification of tris(chlorophenyl)methanol in blubber of harbour seals from Puget Sound. *Chemosphere*, 18: 1799–1804.

Warburton, J. and D.J. Seagars. 1993. *Heavy Metal Concentrations in Liver and Kidney Tissues of Pacific Walrus: Continuation of a Baseline Study*. US Fish and Wildlife Service Technical Report R7/MMM/93-1.

Watanabe, S., T. Shimada, S. Nakamura, N. Nishiyama, N. Yamashita, S. Tanabe, and R. Tatsukawa. 1989. Specific profile of liver microsomal cytochrome P-450 in dolphin and whales. *Marine Environmental Research*, 27: 51–65.

Wolman, A.A. and A.J. Wilson. 1970. Occurrence of pesticides in whales. *Pesticides Monitoring Journal*, 4: 8–10.

Young, D., M. Becerra, D. Kopec, and S. Echols. 1998. GC/MS analysis of PCB congeners in blood of the harbor seal *Phoca vitulina* from San Francisco Bay. *Chemosphere*, 37: 711–733.

Zeisler, R., R.Demiralp, B.J. Koster, P.R. Becker, M. Burow, P. Ostapczuk, and S.A. Wise. 1993. Determination of inorganic constituents in marine mammal tissues. *The Science of the Total Environment*, 139/140: 365–386.

10

Persistent organic pollutants and outbreaks of disease in marine mammals

Peter S. Ross

Introduction

Marine mammals have become unwitting sentinels of the POP contamination of the world's oceans. Seals, whales, and dolphins are often situated at a high position in marine food chains, and are therefore predisposed to accumulating many types of fat-soluble toxic chemicals. There are many different species of marine mammals, which vary in form, ecology, and behaviour. These animals can be broadly divided into the pinnipeds (seals, sea-lions, fur seals, and the walrus, *Odobenus rosmarus*), cetaceans (whales, dolphins, and porpoises), and sirenians (dugongs and manatees), although polar bears (*Ursis maritimus*) and sea otters (*Enhydra lutris*) are often included in this category because of their semi-aquatic existence and their consumption of marine prey species. Each species of marine mammal has evolved special features and strategies that allow them to capture their preferred prey items. Food items vary from large quantities of tiny krill in the case of the blue whale (*Balaenoptera musculus*) to herring (*Clupea harengus*) and hake (*Merluccius productus*) for the temperate harbour seal (*Phoca vitulina*), and even marine mammals in the case of certain populations of killer whales (*Orcinus orca*).

From an environmental pollution perspective, the dietary preferences of these animals is important, since exposure via the consumption of prey represents the route by which mammals accumulate the fat-soluble chemicals. Organisms occupying high positions in marine food chains, including fish-eating birds, pinnipeds, and cetaceans, have been found to have high levels of POPs in many parts of the world (Muir *et al.,* 1996; Ross *et al.,* 1996a). The baleen whales, which occupy lower positions in the food chain, have been found to be much less contaminated by the same chemicals (Gauthier, Metcalfe, and Sears, 1997; Jarman *et al.,* 1996). The herbivorous manatee, *Trichechus manatus,* is also relatively uncontaminated (Ames and Van Vleet, 1996). In some species, different dietary preferences have resulted in certain sub-populations of animals having considerably higher concentrations of POPs than other animals of the same species (Muir *et al.,* 1995; Ross *et al.,* 2000a).

POPs are contaminants of particular concern, because they are resistant to breakdown, persistent in the environment, fat-soluble, and often toxic. These include dichlorodiphenyl trichloroethane (DDT), a pesticide used for agricultural purposes and mosquito control; polychlorinated biphenyls (PCBs), an industrial oil used as a lubricant and heat-resistant stabilizer in various industrial processes and electrical transformers; and polychlorinated dibenzo-*p*-dioxins (PCDDs) and furans (PCDFs), by-products of chlorine bleaching operations at pulp mills, low-temperature combustion processes, and pesticide production processes. Their physico-chemical characteristics lead to a stepwise accumulation of many of these contaminants at increasing concentrations up through the food chain, from minute bacteria and single-celled algae to the fish-eating seals and whales. The ability of biota, including marine mammals, to eliminate certain POPs also influences the pattern of bioaccumulation in aquatic food chains, with the more chlorinated (more fat-soluble) contaminants being more resistant to metabolic processes (Boon *et al.,* 1997; Tanabe *et al.,* 1988).

By the 1960s and early 1970s, it had become apparent that some of the "miracle chemicals" designed and produced for the control of malaria, weeds, and pests, and certain industrial chemicals, were widespread in the environment, affecting some wildlife, and considered a human health threat (Carson, 1962). Despite increased regulations that eliminated PCBs and DDT from usage throughout the industrialized world in the early to mid-1970s, these classes of chemicals have proven very persistent in the environment. Environmental contaminants including the POPs have been distributed around the world through atmospheric processes, deposited on to water surfaces, and have subsequently been incorporated into marine food chains (Hargrave *et al.,* 1988; Patton *et al.,* 1989; Wilkening, Barrie, and Engle, 2000). Antarctic leopard seals, *Hydrurga leptonyx,* were found to contain *p,p'*-DDT and its breakdown product, *p,p'*-DDE, as well as low levels of PCBs in blubber samples collected in 1975 (Risebrough *et al.,* 1976). At the opposite end of the globe, blubber samples collected from ringed seals, *Phoca hispida,* in the Canadian Arctic in 1972 contained PCBs and DDT (Addison and Smith, 1998). More recently, the largely salmon-eating killer whales (*Orcinus orca*) of British Columbia, Canada, have been found to be among the most PCB-contaminated marine mammals in the world, reflecting a combination of local and global contamination (Ross *et al.,* 2000a).

While such studies demonstrated that POPs could be detected in marine mammals far removed from areas of human activity, proximity to industry and agriculture is important. The marine mammal populations of the Baltic and North Seas in northern Europe, the St Lawrence estuary in eastern Canada, and the coastal waters of California have been found to be particularly contaminated with POPs including PCBs and DDT (Blomkvist *et al.*, 1992; Delong, Gilmartin, and Simpson, 1973; Martineau *et al.*, 1987; Muir *et al.*, 1996).

While an increased incidence of disease was associated with high levels of POPs in California sea-lions in 1970 (Delong, Gilmartin, and Simpson, 1973; Gilmartin *et al.*, 1976), it was the mass mortality of harbour and grey (*Halichoerus grypus*) seals in northern Europe in 1988 (reviewed extensively in Dietz, Heide-Jörgensen, and Härkönen, 1989; Hall, Pomeroy, and Harwood, 1992; Osterhaus *et al.*, 1995) which drew widespread attention to the immuno-toxic risk that PCBs and other chemicals presented to marine mammals. Elevated concentrations of POPs were increasingly suspected of causing adverse effects in marine mammals, including a reduction in their ability to defend themselves against the onslaught of pathogens.

Further virus-associated mass mortalities among pinniped and cetacean species heightened interest in the health of marine mammal populations in different parts of the world. Viruses were identified as responsible for several outbreaks, but con-taminants were also suspected by many to have contributed to the severity of the events. The multidisciplinary nature of the issue was further complicated by interactions among a mixture of many hundreds of different contaminants in affected animals; the dynamic, interactive, and sometimes redundant components of the immune system of individuals within a population; and the introduction and spread of a new pathogen (e.g. virus) in and among naïve populations of animals.

Contaminant-associated effects in free-ranging marine mammals

There exists little in the way of definitive evidence that ambient concentrations of environmental contaminants have affected free-ranging marine mammals. This is largely because of the tremendous challenges that are associated with scientific studies of marine mammals, and the complex nature of the toxic responses being evaluated. However, a growing body of literature suggests that populations of seals inhabiting the contaminated coastal waters of northern Europe, and cetaceans frequenting the highly polluted St Lawrence estuary in eastern Canada and the Mediterranean Sea, are being impacted in different ways. Evidence began to implicate POPs in reproductive impairment and skeletal abnormalities among harbour, grey, and ringed seals of the Baltic and Wadden Seas (Bergman, Olsson, and Reiland, 1992; Helle, Olsson, and Jensen, 1976a, 1976b; Mortensen *et al.*, 1992; Reijnders, 1980; Zakharov and Yablokov, 1990), and tumours and low reproductive success among St Lawrence beluga whales (*Delphinapterus leucas*)

(Béland *et al.*, 1993; De Guise *et al.*, 1995). This evidence was generated by studies of stranded (deceased) animals and archived specimens, as well as observational studies of the populations in question.

While the evidence has often appeared circumstantial, such observations have been supported by more mechanistic studies using captive animals under carefully controlled conditions. A captive feeding study of harbour seals, fed either relatively uncontaminated eastern Wadden Sea fish or fish from the more contaminated south-western Wadden Sea, provided the first direct evidence that contaminants could affect reproduction (Reijnders, 1986). Diminished vitamin A and thyroid hormone levels in plasma were also observed in the seals that were fed the contaminated fish, consistent with effects observed in laboratory animals exposed to some of the POPs (Brouwer, Reijnders, and Koeman, 1989). Prior to the 1988 mass mortality of harbour and grey seals in Europe, virtually nothing was known about the immune system of marine mammals, or the effects of chemical contaminants on the immune system of these animals.

Chemicals and the immune system

The immune system of mammals consists of a highly complex set of organs, tissues, cells, and soluble factors, which act in concert to defend the host against invasion by viruses, bacteria, macroparasites, and tumours. The structure and function of the immune system appear to be highly conserved among mammalian species, with pinnipeds and cetaceans having similar white blood cell (leukocyte) sub-populations, lymphoid organs, and immune responses to other mammals studied (Cavagnolo, 1979; De Guise *et al.*, 1997; De Swart *et al.*, 1993; Ross *et al.*, 1993, 1994). The immune response is complex, but can be considered in the simple functional terms of *non-specific immunity*, involving aspects of the immune system that act as a first line of defence against pathogens, and *specific immunity*, the highly effective, memory-based defence against pathogens previously encountered. In both types of responses, different sub-populations of leukocytes circulate in the blood and through lymphoid organs, responding in concert with a number of other cell types, chemical messengers, and soluble factors to eliminate pathogens from the system.

It has long been known that many POPs are toxic to the mammalian immune system. Of the chemicals tested in studies of laboratory animals, the dioxin-like compounds (e.g., PCBs, PCDDs, and PCDFs) appear to be particularly toxic to the immune system. Laboratory animals exposed to these classes of compounds have exhibited thymus atrophy, and impaired T-cell and delayed-type hypersensitivity responses (De Heer *et al.*, 1994; Vos and Van Driel-Grootenhuis, 1972). Contaminant-related impairment of immune responses in laboratory animals leads to a reduced capacity of the host to clear pathogens, including bacteria, viruses, and parasites (Loose *et al.*, 1978; Luebke *et al.*, 1994; Thomas *et al.*,

1987). The link between exposure to contaminants, immunotoxicity, and increased susceptibility to disease has therefore been clearly elucidated in studies of laboratory animals.

The identification of the cytosolic Aryl hydrocarbon receptor (A*h*R) as an intracellular target for the highly toxic 2,3,7,8-tetrachlorodibenzo-*p*-dioxin (TCDD) in mammalian cells has provided a basis for much work on the toxicity of "dioxin-like" compounds, as well as a means for comparative toxicological studies in different species. Chlorinated chemicals that consist of two phenyl rings and assume a planar form, similar to that of 2,3,7,8-TCDD, can bind to varying degrees to the A*h*R and initiate a series of toxic responses (Safe, 1992, 1997). While 2,3,7,8-TCDD represents that most toxic of these chemicals, many of the 74 other PCDDs, 135 PCDFs, and 209 PCBs have been shown to bind to the A*h*R and cause "dioxin-like" toxicities (Table 10.1). The immune system has been identified as a sensitive target for this class of chemicals, with A*h*R-mediated atrophy of the thymus and subsequent impairment of T-lymphocyte function being characteristic of "dioxin-like" immunotoxicity in different animal species (Greenlee *et al.*, 1985; Silkworth and Grabstein, 1982; Thomas and Hinsdill, 1979). As shown in Table 10.1, there exist up to 209 different PCBs, 75 PCDDs and 135 PCDFs in environmental mixtures. Understanding the toxicity of each congener is a seemingly impossible task, but the development of the "toxic equivalents" (TEQ) approach to the most toxic of these 419 congeners, 2,3,7,8-TCDD, has created an integrated way to characterize the toxicity of complex mixtures. By multiplying the concentration measured in a sample by the toxic equivalency factor (TEF), and summing the resulting values for each of these congeners, a total "dioxin equivalent" provides researchers and regulators with a simplified means of assessing the "dioxin-like' toxicity of a sample (TEFs from Van den Berg *et al.*, 1998).

Table 10.1 The "dioxin-like" activities of planar PCBs, PCDDs, and PCDFs act via a common mechanism of action in mammals

PCBs (IUPAC)	TEF	PCDDs	TEF	PCDFs	TEF
81	0.0001	2378-TCDD	1	2378-TCDF	0.1
77	0.0001	12378-PeCDD	1	12378-PeCDF	0.05
126	0.1	123478-HxCDD	0.1	23478-PeCDF	0.5
169	0.01	123678-HxCDD	0.1	123478-HxCDF	0.1
105	0.0001	123789-HxCDD	0.1	123678-HxCDF	0.1
114	0.0005	1234678-HpCDD	0.01	123789-HxCDF	0.1
118	0.0001	OCDD	0.0001	234678-HxCDF	0.1
123	0.0001			1234678-HpCDF	0.01
156	0.0005			1234789-HpCDF	0.01
157	0.0005			OCDF	0.0001
167	0.00001				
189	0.0001				

While contaminants can upset the balance of the immune system, it is important to recognize that many other factors can affect the competence of the immune system in animals, and can play important roles in outbreaks of disease in wildlife populations. Many conditions can and do affect immune function in mammals, such as age, malnutrition, stress, and reproductive cycles (Chandra and Kumari, 1994; Irwin, 1994; Ross et al., 1994). Such factors are likely play a role in the course of disease outbreaks among populations of marine mammals (Hall, Pomeroy, and Harwood, 1992), and should be considered when examining the possible involvement of immunotoxic contaminants.

Disease outbreaks in marine mammals

Virus and bacterial infections represent normal occurrences in wildlife. Most diseases can be expected to result in short-term subclinical or clinical manifestations that disappear following immunological clearance of the causative pathogen. However, occasional episodes of highly infectious and potentially fatal diseases also occur in free-ranging animal populations. Such dramatic events can result from the introduction of a new pathogen to an immunologically naïve population. Several large-scale mortalities have taken place among marine mammals in recent years, with morbilliviruses often being identified as causal agents (De Swart et al., 1995a; Osterhaus et al., 1995).

Very high levels of PCBs and DDT were found in female California sea-lions, *Zalophus californianus*, which were suffering from a high incidence of abortions and premature pupping (Delong, Gilmartin, and Simpson, 1973; Gilmartin et al., 1976). Two infectious agents (the bacterium *Leptospira pneumona*, and the newly identified San Miguel sea-lion virus) were isolated from some of the victims, and may have been responsible for the reproductive effects observed in the population. While it was not possible to establish a causal link between contaminants and the reproductive failures, DDT has been associated with reproductive impairment in birds (Hickey and Anderson, 1968; Lundholm, 1997), and PCBs are immunotoxicants (Vos and Van Driel-Grootenhuis, 1972).

While links between contaminants and the observations mentioned above served to focus scientific efforts, it was not until the 1988 mass mortality of harbour and grey seals in northern Europe that the public and scientific communities were galvanized into action. Many of the 20,000 harbour seals and several hundred grey seals that died suffered from a myriad of symptoms (Hall, Pomeroy, and Harwood, 1992), prompting observers to ponder openly about the fate of the North Sea ecosystem (Dietz, Heide-Jörgensen, and Härkönen, 1989). In the end, a newly identified member of the genus *Morbillivirus* (phocine distemper virus, or PDV) was isolated (Osterhaus and Vedder, 1988), and the answer appeared simple enough: thousands of seals died because they were exposed to a new and highly infectious virus. However, evidence that a similar virus had infected other populations of harbour seals with no record of apparent mortality in less contaminated parts of the world (Henderson *et al.*, 1992; Ross *et al.*, 1992) fuelled speculation that the European disease outbreak was unusual and that additional environmental factors may have contributed to the event.

At approximately the same time, the deaths of approximately 10,000 Baikal seals (*Phoca sibirica*) were attributed to another *Morbillivirus* normally found in dogs (canine distemper virus, or CDV; Grachev *et al.*, 1989; Visser *et al.*, 1990). High PCB concentrations found in the blubber of Baikal seals in later studies added to the speculation that contaminants could be playing a role in *Morbillivirus* outbreaks (Nakata *et al.*, 1995).

The deaths of more than 1,500 striped dolphins (*Stenella coeruleoalba*) in the Mediterranean Sea between 1990 and 1993 resulted from infection with a previously unidentified *Morbillivirus* (dolphin morbillivirus, or DMV; Aguilar and Raga, 1993; Domingo *et al.*, 1990; Van Bressem *et al.*, 1991, 1993). Blubber samples taken from victims of this disease outbreak were found to be heavily contaminated with PCBs compared to biopsy samples taken from free-ranging dolphins in years prior and subsequent to the mortality event, highlighting a possible role for contaminants in the outbreak (Aguilar and Borrell, 1994; Kannan *et al.*, 1993).

Approximately 50 per cent of the inshore population of bottlenose dolphins, *Tursiops truncatus*, off the eastern seaboard of North America succumbed to a *Morbillivirus* infection in 1987 (Lipscomb *et al.*, 1994). High concentrations of PCBs were found in this species, and the authors considered these immunotoxic chemicals to have contributed to the event (Kuehl, Haebler, and Potter, 1991).

While nothing is known of contaminants in a localized population of crabeater seals that suffered a die-off near an Antarctic research station in 1955 (Laws and Taylor, 1957), it should be noted that *Morbillivirus* outbreaks can cause periodic incidents of serious disease in the absence of contaminants (Harder *et al.*, 1995).

However, the high concentrations of immunotoxic chemicals found in the victims of most of these outbreaks suggested a hypothetical involvement of contaminants in these outbreaks that was consistent with laboratory animal studies. Proving or disproving a link between contaminants and the integrity of the immune systems of marine mammals during such outbreaks is not possible. During episodes of disease, the immune response affects the activities of leukocyte sub-populations as they respond to the infection, or is affected by the immunosuppressive nature of certain pathogens. A wide variation in immune function can therefore be expected in different animals in the population. The time course and degree of infection, genetic variation, and the age and condition of individuals will vary within a population.

In addition, functional tests for immune function involve the live culture of leukocyte sub-populations in co-incubation with stimulatory compounds (mitogens) over a period of days (lymphocyte proliferation assays) or in the presence of labelled target cells (cytotoxic cell assays). The results of tests of immune function are generally expressed relative to each other, or to suitable controls.

In this manner, a meaningful assessment of the influences of environmental contaminants requires that samples be obtained from all study animals within hours of each other, and that leukocyte fractions isolated from these samples be assayed concurrently for immune function. Captive or semi-field study designs therefore offer tremendous advantages to the study of immunotoxicology in marine mammals. They allow for groups of same-age animals under similar conditions to be sampled, and also allow for blood samples to be taken rapidly under highly controlled conditions.

Shortly after the PDV-associated mass mortality among seals in northern Europe, a captive feeding study was undertaken to assess the effects of chemical contaminants on immune function. Two groups of 11 harbour seals each were fed herring from either the relatively uncontaminated Atlantic Ocean or the highly contaminated Baltic Sea during a period of 30 months. Animals were caught as weaned pups, and were therefore the same age. There was an equal distribution of males and females in the two groups. After an acclimation period of one year on the same diet, the 22 seals were divided into two groups and fed the respective study diets. Blood samples were drawn every six to nine weeks, and a series of immune function tests was carried out.

Within one year of the start of the feeding study, *in vitro* natural killer cell activity (Ross *et al.*, 1996b) and T-cell function (De Swart *et al.*, 1994, 1995b) were significantly lower in the group of seals fed the more contaminated diet of Baltic Sea herring. Two years after the start of the study, delayed-type hypersensitivity (DTH) responses to ovalbumin, a measure of T-cell function *in vivo*, was lower in the group of seals fed the contaminated herring (Ross *et al.*, 1995). B-cell function was largely unaffected in *in vitro* tests (De Swart *et al.*, 1994).

The pattern of observations in the seals was consistent with observations of laboratory animals exposed to dioxin-like compounds. Subsequent studies by the same researchers examined the effects of the Atlantic Ocean and Baltic Sea herring on immune function and host resistance in laboratory rats. They found that T-cell function and NK-cell activity were adversely affected in rats exposed to Baltic Sea herring through diet or perinatally (Ross *et al.*, 1996c, 1997). In addition, these studies revealed an effect of contaminants from the Baltic Sea herring on the weight and cellularity of the thymus, suggesting that the thymus may have been an important target for dioxin-like contaminants in both the rat and the seal studies.

The profile of contaminants in blubber biopsies taken from the captive seals in these studies revealed that over 90 per cent of the total TEQ to 2,3,7,8-TCDD were due to PCBs, while dioxins and furans contributed only small amounts. Given the dioxin-like pattern of immune function effects observed in both the rats and the seals, the authors concluded that the PCBs were largely responsible for the immunotoxicity in the captive study of harbour seals. Since many populations of harbour seals and other marine mammal species have concentrations of PCBs that exceed the accumulated levels found in biopsies of the captive Baltic group of seals (16.8 mg/kg lipid), the authors concluded that environmental contaminants could be affecting marine mammals in the oceans (Figure 10.1; Ross *et al.*, 1996a).

Captive studies provide the best means of controlling for confounding factors and ensuring the comparability of immune function results among individual marine mammals. However, immunotoxicological studies of free-ranging marine mammals have been carried out. Although sample size was limited, blood samples from five free-ranging bottlenose dolphins yielded negative correlations between T-cell function and PCB and DDT concentrations (Lahvis *et al.*, 1995). Studies combining the *in vitro* culture of leukocytes isolated from the blood of beluga whales with PCBs and metals found that the function of these white blood cells was impaired at concentrations encountered in free-ranging whales (De Guise *et al.*, 1998). Such studies provide additional evidence that concentrations of environmental contaminants in marine mammals are high enough to elicit immunotoxic effects in the natural environment.

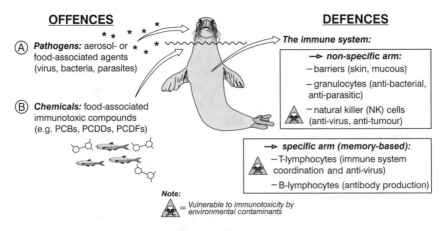

Figure 10.1 Persistent organic pollutants (POPs) can affect the health of marine mammals in the world's oceans by impairing their immune systems.

Fish-eating marine mammals often occupy high positions in ocean food chains. As a result, they are predisposed to accumulating high concentration of POPs, such as the polychlorinated biphenyls (PCBs). Infectious agents, including viruses and bacteria (A), constantly challenge healthy marine mammal populations, but when immunotoxic contaminants (B) accumulate to high concentrations in the food chain, the immune system (on the right of the diagram) may be incapable of responding effectively to the assault of these agents. Captive studies have shown that elements of both the non-specific immune system (natural killer cells) and the specific immune system (T cells) in harbour seals are affected by contaminants (triangle with "skull and crossbones" symbol).

Contaminant trends: marine mammals in the twenty-first century

The implementation of more stringent regulation of PCBs and DDT led to declines in the concentrations of these contaminants in various environmental compartments through the 1970s and into the 1980s (Andersson *et al.*, 1988), with concurrent declines being observed in marine mammals as well (Addison and Smith, 1998; Muir *et al.*, 1988; Tanabe *et al.*, 1994). However, little change was observed in PCB and DDT levels from the mid-1980s through the late 1990s, suggesting that ongoing seepage from dumps and recycling in the environment is contributing to a levelling-off of contaminant levels (Addison and Smith, 1998; Tanabe *et al.*, 1994). The contribution of persistent PCBs to the total TEQ in a number of populations in the world suggest that these compounds will continue to present an immunotoxic risk to marine mammals for some time (Ross *et al.*, 1996a).

While episodes of infectious disease in marine mammal populations are no doubt due to a combination of biological and ecological factors, PCBs and related compounds will likely contribute to the severity of such outbreaks in the more contaminated populations during the coming decades. Infectious diseases will probably present an increasing threat to the world's marine mammals as a consequence of the effects of POPs and other anthropogenic factors on the occurrence, frequency, and transmission of pathogens within and across species (Daszak, Cunningham, and Hyatt, 2000; Harvell *et al.*, 1999; Mahy and Brown, 2000; Ross, 2002). Contaminant-associated outbreaks of infectious disease may not always be as visible as the 1988 mortality of 20,000 harbour and grey seals in Europe. There have been numerous reports of pathogen-related mortalities among newborn marine mammals (Baker, 1989; Garner *et al.*, 1997), and the elusive nature of marine mammals in the oceans render it impossible to detect all victims, quantify precise mortality rates in a population, or obtain samples for study. Subtle, yet ecologically significant, influences of POPs on disease events in free-ranging populations may take place in ways that are difficult to measure.

Future immunotoxicological studies must minimize the many confounding factors that can affect immune function results in marine mammals. While it will remain extremely difficult to ascertain the ecological significance of the results of immunotoxicological studies, policy-makers should take advantage of the weight of evidence available from the combined results of laboratory animal studies, captive marine mammal studies, and studies of free-ranging marine mammals (Ross *et al.*, 2000b; Ross, 2000). Use of a precautionary approach in ecological risk assessment for issues of POPs in marine mammal populations is currently the best available strategy for regulators and conservationists. Given the often elevated trophic level of marine mammals in the world's oceans, the information generated from high-quality studies can provide us with an integrated "state of the marine environment" report.

Marine mammals can serve as valuable "sentinel" species, and will hopefully help humans to understand their impact on the world's oceans. While many of the tradition or "legacy" POPs have been regulated in the industrial world, some continue to be used in developing nations. In addition, many of the newer replacement compounds share problematic features with the regulated POPs: they are persistent, bioaccumulative, and toxic. Given the rapidity at which such contaminants can move around the world through atmospheric processes, the oceans will continue to serve as sinks for anthropogenic contaminants.

REFERENCES

Addison, R.F. and T.G. Smith. 1998. Trends in organochlorine residue concentrations in ringed seal (*Phoca hispida*) from Holman, Northwest Territories, 1972–1991. *Arctic*, 51: 253–261.

Aguilar, A. and A. Borrell. 1994. Abnormally high polychlorinated biphenyl levels in striped dolphins (*Stenella coeruleoalba*) affected by the 1990–1992 Mediterranean epizootic. *Science of the Total Environment*, No. 154, 237–247.

Aguilar, A. and J.A. Raga. 1993. The striped dolphin epizootic in the Mediterranean Sea. *AMBIO*, 22: 524–528.

Ames, A.L. and E.S. Van Vleet. 1996. Organochlorine residues in the Florida manatee, *Trichechus manatus latirostris*. *Marine Pollution Bulletin*, 32: 374–377.

Andersson, O., C.-E. Linder, M. Olsson, L. Reutergårdh, U.-B. Uvemo, and U. Wideqvist. 1988. Spatial differences and temporal trends of organochlorine compounds in biota from the north-western hemisphere. *Archives of Environmental Contamination and Toxicology,*17: 755–765.

Baker, J.R. 1989. Natural causes of death in non-suckling grey seals (*Halichoerus grypus*). *Veterinary Record*, 125: 500–503.

Béland, P., S. De Guise, C. Girard, A. Lagacé, D. Martineau, R. Michaud, D.C.G. Muir, R.J. Norstrom, E. Pelletier, S. Ray, and L.R. Shugart. 1993. Toxic compounds and health and reproductive effects in St Lawrence beluga whales. *Journal of Great Lakes Research*, 19: 766–775.

Bergman, A., M. Olsson, and S. Reiland. 1992. Skull-bone lesions in the Baltic grey seal (*Halichoerus grypus*). *AMBIO*, 21: 517–519.

Blomkvist, G., A. Roos, S. Jensen, A. Bignert, and M. Olsson. 1992. Concentrations of sDDT and PCB in seals from Swedish and Scottish waters. *AMBIO*, 21: 539–545.

Boon, J.P., J. Van der Meer, C.R. Allchin, R.J. Law, J. Klungsoyr, P.E.G. Leonards, H. Spliid, E. Storr-Hansen, C. Mckenzie, and D.E. Wells. 1997. Concentration-dependent changes of PCB patterns in fish-eating mammals: Structural evidence for induction of cytochrome P450. *Archives of Environmental Contamination and Toxicology*, No. 33, 298–311.

Brouwer, A., P.J.H. Reijnders, and J.H. Koeman. 1989. Polychlorinated biphenyl (PCB)-contaminated fish induces vitamin A and thyroid hormone deficiency in the common seal (*Phoca vitulina*). *Aquatic Toxicology*, No. 15, 99–106.

Carson, R. 1962. *Silent Spring*, Boston, MA: Houghton-Mifflin, p. 368.

Cavagnolo, R.Z. 1979. The immunology of marine mammals. *Developmental and Comparative Immunology*, No. 3, 245–257.

Chandra, R.K. and S. Kumari. 1994. Nutrition and immunity: an overview. *Journal of Nutrition*, No. 124, 1433S–1435S.

Daszak, P., A.A. Cunningham, and A.D. Hyatt. 2000. Emerging infectious diseases of wildlife – threats to biodiversity and human health. *Science*, 287: 443–449.

De Guise, S., D. Martineau, P. Béland, and M. Fournier. 1995. Possible mechanisms of action of environmental contaminants on St. Lawrence beluga whales (*Delphinapterus leucas*). *Environmental Health Perspectives Supplements*, No. 103, 73–77.

De Guise, S., D. Martineau, P. Béland, and M. Fournier. 1998. Effects of *in vitro* exposure of beluga whale leukocytes to selected organochlorines. *Journal of Toxicology and Environmental Health*, No. 55, 479–493.

De Guise, S., P.S. Ross, A.D.M.E. Osterhaus, D. Martineau, P. Béland, and M. Fournier. 1997. Immune functions in beluga whales (*Delphinapterus leucas*): evaluation of natural killer cell activity. *Veterinary Immunology and Immunopathology*, 58: 345–354.

De Heer, C., A.P.J. Verlaan, A.H. Penninks, J.G. Vos, H.J. Schuurman, and H. Van Loveren. 1994. Time course of 2,3,7,8-tetrachlorodibenzo-*p*-dioxin (TCDD)-induced thymic atrophy in the Wistar rat. *Toxicology and Applied Pharmacology*, 128: 97–104.

Delong, R.L., W.G. Gilmartin, and J.G. Simpson. 1973. Premature births in California sea lions: Association with high organochlorine pollutant residue levels. *Science*, No. 181, 1168–1170.

De Swart, R.L., R.M.G. Kluten, C.J. Huizing, L.J. Vedder, P.J.H. Reijnders, I.K.G. Visser, F.G.C.M. UytdeHaag, and A.D.M.E. Osterhaus. 1993. Mitogen and antigen induced B and T cell responses of peripheral blood mononuclear cells from the harbour seal (*Phoca vitulina*). *Veterinary Immunology and Immunopathology*, No. 37, 217–230.

De Swart, R.L., T.C. Harder, P.S. Ross, H.W. Vos, and A.D.M.E. Osterhaus. 1995a. Morbilliviruses and morbillivirus diseases of marine mammals. *Infectious Agents and Disease*, 4: 125–130.

De Swart, R.L., P.S. Ross, H.H. Timmerman, H.W. Vos, P.J.H. Reijnders, J.G. Vos, and A.D.M.E. Osterhaus. 1995b. Impaired cellular immune response in harbour seals (*Phoca vitulina*) feeding on environmentally contaminated herring. *Clinical and Experimental Immunology*, 101: 480–486.

De Swart, R.L., P.S. Ross, L.J. Vedder, H.H. Timmerman, S.H. Heisterkamp, H. Van Loveren, J.G. Vos, P.J.H. Reijnders, and A.D.M.E. Osterhaus. 1994. Impairment of immune function in harbor seals (*Phoca vitulina*) feeding on fish from polluted waters. *Ambio*, 23: 155–159.

Dietz, R., M.-P. Heide-Jörgensen, and T. Härkönen. 1989. Mass deaths of harbor seals (*Phoca vitulina*) in Europe. *AMBIO*, 18: 258–264.

Domingo, M., L. Ferrer, M. Pumarola, A. Marco, J. Plana, S. Kennedy, M. McAlisky, and B.K. Rima. 1990. Morbillivirus in dolphins. *Nature*, No. 348, 21.

Garner, M.M., D.M. Lambourn, S.J. Jeffries, P.B. Hall, J.C. Rhyan, D.R. Ewalt, L.M. Polzin, and N.F. Cheville. 1997. Evidence of *Brucella* infection in Parafilaroides lungworms in a Pacific harbor seal (*Phoca vitulina richardsi*). *Journal of Veterinary Diagnostic Investigation*, No. 9, 298–303.

Gauthier, J.M., C.D. Metcalfe, and R. Sears. 1997. Chlorinated organic contaminants in blubber biopsies North-western Atlantic balaenopterid whales summering in the Gulf of St. Lawrence. *Marine Environmental Research*, No. 44, 201–223.

Gilmartin, W.G., R.L. Delong, A.W. Smith, J.C. Sweeney, B.W. De Lappe, R.W. Risebrough, L.A. Griner, M.D. Dailey, and D.B. Peakall. 1976. Premature parturition in the California sea lion. *Journal of Wildlife Diseases*, No. 12, 104–115.

Grachev, M.A., V.P. Kumarev, L.V. Mamaev, V.L. Zorin, L.V. Baranova, N.N. Denikjna, S.I. Belikov, E.A. Petrov, V.S. Kolesnik, R.S. Kolesnik, V.M. Dorofeev, A.M. Beim, V.N. Kudelin, F.G. Nagieva, and V.N. Sidorov. 1989. Distemper virus in Baikal seals. *Nature*, No. 338, 209.

Greenlee, W.F., K.M. Dold, R.D. Irons, and R. Osborne. 1985. Evidence for direct action of 2,3,7,8-tetrachlorodibenzo-p-dioxin (TCDD) on thymic epithelium. *Toxicology and Applied Pharmacology*, No. 79, 112–120.

Hall, A.J., P.P. Pomeroy, and J. Harwood. 1992. The descriptive epizootiology of phocine distemper in the UK during 1988/89. *Science of the Total Environment*, No. 115, 31–44.

Harder, T.C., M. Kenter, M.J.G. Appel, M.E. Roelke-Parker, T. Barrett, and A.D.M.E. Osterhaus. 1995. Phylogenetic evidence of canine distemper virus in Serengeti's lions. *Vaccine*, No. 13, 521–523.

Hargrave, B.T., W.P. Vass, P.E. Erickson, and B.R. Fowler. 1988. Atmospheric transport of organochlorines to the Arctic Ocean. *Tellus*, 40B: 480–493.

Harvell, C.D., K. Kim, J.M. Burkholder, R.R. Colwell, P.R. Epstein, D.J. Grimes, E.E. Hofmann, E.K. Lipp, A.D.M.E. Osterhaus, R.M. Overstreet, J.W. Porter, G.W. Smith, and G.R. Vasta. 1999. Emerging marine diseases – climate links and anthropogenic factors. *Science*, 285: 1505–1510.

Helle, E., M. Olsson, and S. Jensen. 1976a. PCB levels correlated with pathological changes in seal uteri. *AMBIO*, 5: 261–263.

Helle, E., M. Olsson, and S. Jensen. 1976b. DDT and PCB levels and reproduction in ringed seal from the Bothnian Bay. *Ambio*, 5: 188–189.

Henderson, G., A. Trudgett, C. Lyons, and K. Ronald. 1992. Demonstration of antibodies in archival sera from Canadian seals reactive with a European isolate of phocine distemper virus. *Science of the Total Environment*, No. 115, 93–98.

Hickey, J.J. and D.W. Anderson. 1968. Chlorinated hydrocarbons and eggshell changes in raptorial and fish-eating birds. *Science*, No. 162, 271–273.

Irwin, M. 1994. Stress-induced immune suppression – role of brain corticotropin releasing hormone and autonomic nervous system mechanisms. *Advances in Neuroimmunology*, No. 4, 29–47.

Jarman, W.M., R.J. Nostrom, D.C.G. Muir, B. Rosenberg, M. Simon, and R.W. Baird. 1996. Levels of organochlorine compounds, including PCDDs and PCDFs, in the blubber of cetaceans from the West coast of North America. *Marine Pollution Bulletin*, 32: 426–436.

Kannan, K., S. Tanabe, A. Borrell, A. Aguilar, S. Focardi, and R. Tatsukawa. 1993. Isomer-specific analysis and toxic evaluation of polychlorinated biphenyls in striped dolphins affected by an epizootic in the western Mediterranean Sea. *Archives of Environmental Contamination and Toxicology*, No. 25, 227–233.

Kuehl, D.W., R. Haebler, and C. Potter. 1991. Chemical residues in dolphins from the U.S. Atlantic coast including atlantic bottlenose obtained during the 1987/1988 mass mortality. *Chemosphere*, 22: 1071–1084.

Lahvis, G.P., R.S. Wells, D.W. Kuehl, J.L. Stewart, H.L. Rhinehart, and C.S. Via. 1995. Decreased lymphocyte responses in free-ranging bottlenose dolphins (*Tursiops truncatus*) are associated with increased concentrations of PCBs and DDT in peripheral blood. *Environmental Health Perspectives Supplements*, 103: 67–72.

Laws, R.J. and R.J.F. Taylor. 1957. A mass dying of crabeater seals, *Lobodon carcinophagus* (Gray). *Proceedings of the Zoological Society of London*, No. 129, 315–324.

Lipscomb, T.P., F.Y. Schulman, D. Moffett, and S. Kennedy. 1994. Morbilliviral disease in Atlantic bottlenose dolphins (*Tursiops truncatus*) from the 1987–1988 epizootic. *Journal of Wildlife Diseases*, No. 30, 567–571.

Loose, L.D., J.B. Silkworth, K.A. Pittman, K.-F. Benitz, and W. Mueller. 1978. Impaired host resistance to endotoxin and malaria in polychlorinated biphenyl- and hexachlorobenzene-treated mice. *Infection and Immunity*, 20: 30–35.

Luebke, R.W., C.B. Copeland, J.J. Diliberto, P.I. Akubue, D.L. Andrews, M.M. Riddle, W.C. Williams, and L. Birnbaum. 1994. Assessment of host resistance to *Trichinella spiralis* in mice following preinfection exposure to 2,3,7,8-TCDD. *Toxicology and Applied Pharmacology*, 125: 7–16.

Lundholm, C.E. 1997. DDE-induced eggshell thinning in birds: effects of p,p′-DDE on the calcium and prostaglandin metabolism of the eggshell gland. *Comparative Biochemistry and Physiology B*, 118C: 113–128.

Mahy, B.W.J. and C.C. Brown. 2000. Emerging zoonoses: Crossing the species barrier. *Revue scientifique et technique de l'office international des epizooties*, 19(1): 33–40.

Martineau, D., P. Béland, C. Desjardins, and A. Lagacé. 1987. Levels of organochlorine chemicals in tissues of beluga whales (*Delphinapterus leucas*) from the St. Lawrence Estuary, Québec, Canada. *Archives of Environmental Contamination and Toxicology*, No. 16, 137–147.

Mortensen, P., A. Bergman, A. Bignert, H.-J. Hansen, T. Härkönen, and M. Olsson. 1992. Prevalence of skull lesions in harbor seals (*Phoca vitulina*) in Swedish and Danish museum collections: 1835–1988. *AMBIO*, 21: 520–524.

Muir, D.C.G., C.A. Ford, B. Rosenberg, R.J. Norstrom, M. Simon, and P. Béland. 1996. Persistent organochlorines in beluga whales (*Delphinapterus leucas*) from the St. Lawrence River estuary – Concentrations and patterns of specific PCBs, chlorinated pesticides and polychlorinated dibenzo-*p*-dioxins and dibenzofurans. *Environmental Pollution*, 93: 219–234.

Muir, D.C.G., R.J. Norstrom, and M. Simon. 1988. Organochlorine contaminants in Arctic marine food chains: Accumulation of specific polychlorinated biphenyls and chlordane-related compounds. *Environmental Science and Technology*, 22: 1071–1079.

Muir, D.C.G., M.D. Segstro, K.A. Hobson, C.A. Ford, R.E.A. Stewart, and S. Olpinski. 1995. Can seal eating explain elevated levels of PCBs and organochlorine pesticides in walrus blubber from eastern Hudson Bay (Canada)? *Environmental Pollution*, No. 90, 335–348.

Nakata, H., S. Tanabe, R. Tatsukawa, M. Amano, N. Miyazaki, and E. Petrov. 1995. Persistent organochlorine residues and their accumulation kinetics in Baikal seal (*Phoca sibirica*) from Lake Baikal, Russia. *Environmental Science and Technology*, 29: 2877–2885.

Osterhaus, A.D.M.E. and E.J. Vedder. 1988. Identification of virus causing recent seal deaths. *Nature*, No. 335, 20.

Osterhaus, A.D.M.E., R.L. De Swart, H.W. Vos, P.S. Ross, M.J.H. Kenter, and T. Barrett. 1995. Morbillivirus infections of aquatic mammals: newly identified members of the genus. *Veterinary Microbiology*, No. 44, 219–227.

Patton, G.W., D.A. Hinckley, M.D. Walla, T.F. Bidleman, and B.T. Hargrave. 1989. Airborne organochlorines in the Canadian high arctic. *Tellus*, 41B: 243–255.

Reijnders, P.J.H. 1980. Organochlorine and heavy metal residues in harbour seals from the Wadden Sea and their possible effects on reproduction. *Netherlands Journal of Sea Research*, 14: 30–65.

Reijnders, P.J.H. 1986. Reproductive failure in common seals feeding on fish from polluted coastal waters. *Nature*, 324: 456–457.

Risebrough, R.W., W. Walker, T.T. Schmidt, B.W. De Lappe, and C.W. Connors. 1976. Transfer of chlorinated biphenyls to Antarctica. *Nature*, 264: 738–739.

Ross, P.S. 2000. Marine mammals as sentinels in ecological risk assessment. *Human and Ecological Risk Assessment*, 6: 29–46.

Ross, P.S. 2002. The role of immunotoxic environmental contaminants in facilitating the emergence of infectious diseases in marine mammals. *Human and Ecological Risk Assessment*, 8(2): 277–292.

Ross, P.S., R.L. De Swart, I.K.G. Visser, W. Murk, W.D. Bowen, and A.D.M.E. Osterhaus. 1994. Relative immunocompetence of the newborn harbour seal, *Phoca vitulina*. *Veterinary Immunology and Immunopathology*, 42: 331–348.

Ross, P.S., R.L. De Swart, P.J.H. Reijnders, H. Van Loveren, J.G. Vos, and A.D.M.E. Osterhaus. 1995. Contaminant-related suppression of delayed-type hypersensitivity and antibody responses in harbor seals fed herring from the Baltic Sea. *Environmental Health Perspectives*, 103: 162–167.

Ross, P.S., R.L. De Swart, R.F. Addison, H. Van Loveren, J.G. Vos, and A.D.M.E. Osterhaus. 1996a. Contaminant-induced immunotoxicity in harbour seals: wildlife at risk?. *Toxicology*, 112: 157–169.

Ross, P.S., R.L. De Swart, H.H. Timmerman, P.J.H. Reijnders, J.G. Vos, H. Van Loveren, and A.D.M.E. Osterhaus. 1996b. Suppression of natural killer cell activity in harbour seals (*Phoca vitulina*) fed Baltic Sea herring. *Aquatic Toxicology*, 34: 71–84.

Ross, P.S., R.L. De Swart, H. Van der Vliet, L. Willemsen, A. De Klerk, G. Van Amerongen, J. Groen, A. Brouwer, I. Schipholt, D.C. Morse, H. Van Loveren, A.D.M.E. Osterhaus, and J.G. Vos. 1997. Impaired cellular immune response in rats exposed perinatally to Baltic Sea herring oil or 2,3,7,8-TCDD. *Archives of Toxicology*, 17: 563–574.

Ross, P.S., B. Pohajdak, W.D. Bowen, and R.F. Addison. 1993. Immune function in free-ranging harbor seal (*Phoca vitulina*) mothers and their pups during lactation. *Journal of Wildlife Diseases*, 29: 21–29.

Ross, P.S., H. Van Loveren, R.L. De Swart, H. Van der Vliet, A. De Klerk, H.H. Timmerman, R.S. Van Binnendijk, A. Brouwer, J.G. Vos, and A.D.M.E. Osterhaus. 1996c. Host resistance to rat cytomegalovirus (RCMV) and immune function in adult PVG rats fed herring from the contaminated Baltic Sea. *Archives of Toxicology*, 70: 661–671.

Ross, P.S., I.K.G. Visser, H.W.J. Broeders, M.W.G. Van de Bildt, W.D. Bowen, and A.D.M.E. Osterhaus. 1992. Antibodies to phocine distemper virus in Canadian seals. *Veterinary Record*, 130: 514–516.

Ross, P.S., G.M. Ellis, M.G. Ikonomou, L.G. Barrett-Lennard, and R.F. Addison. 2000a. High PCB concentrations in free-ranging Pacific killer whales, *Orcinus orca*: effects of age, sex and dietary preference. *Marine Pollution Bulletin*, 40: 504–515.

Ross, P.S., J.G. Vos, L.S. Birnbaum, and A.D.M.E. Osterhaus. 2000b. PCBs are a health risk for humans and wildlife. *Science*, 289: 1878–1879.

Safe, S.H. 1992. Development, validation and limitations of toxic equivalency factors. *Chemosphere*, 25: 61–64.

Safe, S.H. 1997. Limitations of the toxic equivalency factor approach for risk assessment of TCDD and related compounds. *Teratog. Carcinog. Mutagen.*, 17: 285–304.

Silkworth, J.B. and E.M. Grabstein. 1982. Polychlorinated biphenyl immunotoxicity: dependence on isomer planarity and the *Ah* gene complex. *Toxicology and Applied Pharmacology*, 65: 109–115.

Tanabe, S., S. Watanabe, H. Kan, and R. Tatsukawa. 1988. Capacity and mode of PCB metabolism in small cetaceans. *Marine Mammal Science*, 4: 103–124.

Tanabe, S., J.-K. Sung, D.-Y. Choi, N. Baba, M. Kiyota, K. Yoshida, and R. Tatsukawa. 1994. Persistent organochlorine residues in northern fur seal from the Pacific Coast of Japan since 1971. *Environmental Pollution*, 85: 305–314.

Thomas, P.T. and R.D. Hinsdill. 1979. The effect of perinatal exposure to tetrachloro-dibenzo-p-dioxin on the immune response of young mice. *Drug and Chemical Toxicology*, 2: 77–98.

Thomas, P.T., H.V. Ratajczak, W.C. Eisenberg, M. Furedi Machacek, K.V. Ketels, and P.W. Barbera. 1987. Evaluation of host resistance and immunity in mice exposed to the carbamate pesticide aldicarb. *Fundamental and Applied Toxicology*, 9: 82–89.

Van Bressem, M.F., I.K.G. Visser, M.W.G. Van de Bildt, J.S. Teppema, J.A. Raga, and A.D.M.E. Osterhaus. 1991. Morbillivirus infection in Mediterranean striped dolphins (*Stenella coeruleoalba*). *Veterinary Record*, 129: 471–472.

Van Bressem, M.F., I.K.G. Visser, R.L. De Swart, C. Örvell, L. Stanzani, E. Androukaki, K. Siakavara, and A.D.M.E. Osterhaus. 1993. Dolphin morbillivirus infection in different parts of the Mediterranean Sea. *Archives of Virology*, 129: 235–242.

Van den Berg, M., L. Birnbaum, A.T.C. Bosveld, B. Brunstrom, P. Cook, M. Feeley, J.P. Giesy, A. Hanberg, R. Hasegawa, S.W. Kennedy, T. Kubiak, J.C. Larsen, F.X.R. Van Leeuwen, A.K. Liem, C. Nolt, R.E. Peterson, L. Poellinger, S.H. Safe, D. Schrenk, D.E. Tillitt, M. Tysklind, M. Younes, F. Waern, and T.R. Zacharewski. 1998. Toxic equivalency factors (TEFs) for PCBs, PCDDs, PCDFs for humans and wildlife. *Environmental Health Perspectives*, 106: 775–792.

Visser, I.K.G., V.P. Kumarev, C. Örvell, P. De Vries, H.W.J. Broeders, M.W.G. Van de Bildt, J. Groen, J.S. Teppema, M.C. Burger, F.G.C.M. UytdeHaag, and A.D.M.E. Osterhaus. 1990. Comparison of two morbilliviruses isolated from seals during outbreaks of distemper in North West Europe and Siberia. *Archives of Virology*, 111: 149–164.

Vos, J.G. and L. Van Driel-Grootenhuis. 1972. PCB-induced suppression of the humoral and cell-mediated immunity in guinea pigs. *Science of the Total Environment*, 1: 289–302.

Wilkening, K.E., L.A. Barrie and M. Engle. 2000. Trans-Pacific air pollution. *Science*, 290: 65–66.

Zakharov, V.M. and A.V. Yablokov. 1990. Skull asymmetry in the Baltic grey seal: effects of environmental pollution. *Ambio*, 19: 266–269.

Part III

Marine biodiversity and environment in the Black Sea and the south-western Atlantic Ocean

11

Biodiversity in the Black Sea: Threats and the future

Bayram Ozturk and Ayaka Amaha Ozturk

Introduction

The Black Sea is one of the world's most isolated seas from the major oceans, and the largest anoxic body of water on the planet (87 per cent of its volume is anoxic). It is surrounded by Bulgaria, Georgia, Romania, Russia, Turkey, and Ukraine (Figure 11.1), where recently socio-economic conditions have been unstable. The total population living in the Black Sea basin (Figure 11.2) is 162 million, while the population along the coasts is approximately 55 million. These figures are enough to show that this sea is under heavy anthropogenic stress. In this chapter, the biodiversity of this unique sea is presented, as well as the factors affecting it. Threats to its biodiversity are then discussed in an attempt to find solutions for the future.

Geological evolution and its influence on biodiversity

The geological evolution of the Black Sea is still controversial (Izdar and Ergun, 1991). However, as it has played an important role for the fauna and flora of this sea, the present authors outline it briefly here, according to Izdar and Ergun (1991) and Zaitsev and Mamaev (1997).

The origin of the Black Sea dates to the Miocene period (5–7 million years ago) when it was the Sarmatic Sea, one of the brackish basins into which the Tethys Sea was divided.

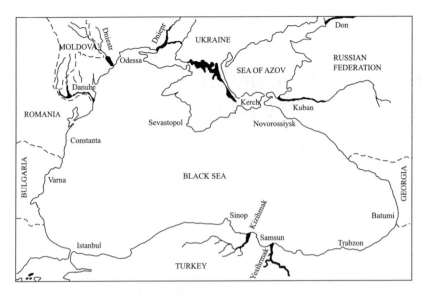

Figure 11.1 The Black Sea and riparian countries

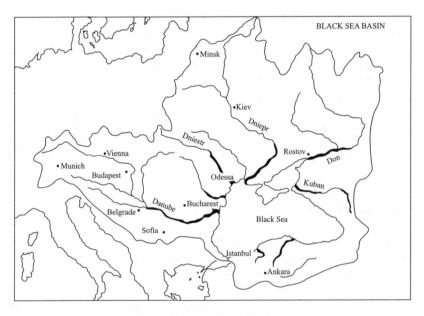

Figure 11.2 The Black Sea basin

In the Early Pliocene (3–5 million years ago), the Sarmatic Sea shrunk to the Maeotic Sea, a body of water whose link with the ocean resulted in increasing salinity, so that marine plants and animals were established. In the Pliocene (1.5–3 million years ago), the connection with the ocean closed again, creating the predominantly freshwater Pontian Sea Lake, within which the present Black and Caspian Seas became connected. Marine fauna were eliminated and brackish-water fauna became dominant. Towards the end of the Pliocene (1 million years ago), the Pontian shrunk to establish the Chaudian Sea Lake, a new body of water separated from the Caspian Sea.

In the late Mindel Glaciation (400,000–500,000 years ago), this Chaudian Sea Lake became filled with fresh water from melting ice and was transformed into the Paleoeuxinian Basin. At this stage, the general geographical outline resembles the present Black and Azov Seas. In the Interglacial Period (100,000–150,000 years ago), the Dardanelles opened and again the connection with the ocean and the Mediterranean started. The so-called Karangat Sea was established, bringing more saline water and representatives of marine fauna and flora (Erinc, 1984). At the end of the last Würm Glaciation (18,000–20,000 years ago), the Karangat was replaced with the Neoeuxinian Lake Sea due to the melting water and again the connection with the ocean was closed. Thus, salinity was greatly reduced and the marine flora and fauna disappeared.

About 7,000–10,000 years ago, the present Black Sea was formed and a connection to the Mediterranean occurred via the Istanbul Strait. A great Mediterraneanization took place and a large number of Mediterranean species settled in it. The Pontian relics moved into the estuaries and bays with low salinity, just as they had during the Karangat period.

Water balance

Due to its geological evolution and isolation from the world's major oceans, the water balance has become an important factor for the marine ecosystems of the Black Sea. The water balance is under the influence of river discharge, precipitation, evaporation, and exchange via the Istanbul and Kerch Straits. It is in equilibrium, although the estimates of inflow and outflow vary among researchers. For example, Unlata et al. (1990) estimated it at 611–612 km^3 per year, while Soljankin (1963; cited in Sorokin, 1983) estimated it at 704 km^3 per year.

The Black Sea's water is renewed from the Mediterranean Sea only through the Turkish Straits system, which includes the Canakkale Strait (Dardanelles), the Marmara Sea, and the Istanbul Strait (Bosphorus). The exchange of water mass through the straits is one of the most influential factors in the hydrological and chemical regime of the Black Sea. The Turkish Straits system carries highly saline water (35 per mill) of Mediterranean origin into the bottom current of the Black Sea, and less saline water (18 per mill) of Black Sea origin into the upper current. The outflow through the Istanbul Strait is twice the inflow. The Istanbul Strait is a narrow and shallow channel whose length is 31 km, minimum width is 740 m, and maximum width is 3.5 km. The depth varies between 30 and 110 m, with an average depth of 50 m.

Another strait related to the water balance of the Black Sea is the Kerch Strait, providing a connection with the Azov Sea and contributing low saline water (5 per mill). The length of this strait is 45 km, and the width varies between 3.7 and 42 km.

The catchment area of the Black Sea is over 2 million km^2, covering 22 countries (Figure 11.2), including six riparian states. International rivers flow into the Black Sea, transporting fresh water and nutrients as well as pollutants. Most important of these rivers are the Danube, Dniestr, Dniepr, Don, Coruh, Kizilirmak, and Yesilirmak. Several highly populated cities – including Munich, Vienna, Budapest, Belgrade, and Kiev – discharge into the Black Sea through these rivers. The prevailing portion of the river runoff is due to the Danube, which is about 2,000 km^3 per year.

Morphometry and oceanographic characteristics

The total surface area of the Black Sea is 423,000 km^2 and the maximum and average depths are 2,212 and 1,240 m respectively. The Black Sea shore is 4,340 km long in total. The Azov Sea, at the north-eastern corner of the Black Sea, has a surface area of 39,000 km^2. Its maximum and average depths are 14 and 8 m, respectively.

The most striking characteristic of the Black Sea is probably the high level of hydrogen sulphide (H_2S). The level of H_2S, which is 150–200 m deep, has been relatively stable (Sorokin, 1983), although seasonal and annual fluctuations have been observed (Oguz, Malanotte-Rizzoli, and Aubrey, 1994). The concentration of dissolved oxygen is 6.7 mg/l at a 10 m depth, 5.0 at 50 m, 0.8 at 100 m, 0.05 at 200 m, and 0.0 in deeper water. On the other hand, the H_2S concentration is 0.15 mg/l at a 150 m depth, 0.75 at 200 m, 5.15 at 500 m, and 9.21 at 2,000 m (Balkas et al., 1990).

The presence of a permanent halocline between 150 and 200 m is another major distinguishing characteristic (Figure 11.3). The stratification is affected by the freshwater input and the Mediterranean inflow of highly saline water. The average surface salinity is about 18–18.5 per mill during winter, and increases by 1.0–1.5 per mill in summer.

The temperature shows more variation than the salinity, on a seasonal as well as on a regional basis. The mean annual surface temperature varies from 16°C in the south to 13°C in the north-east and 11°C in the north-west. While the upper 50–70 m water layer has seasonal fluctuations in temperature, and there is considerable vertical variation, the temperature of the deeper water remains

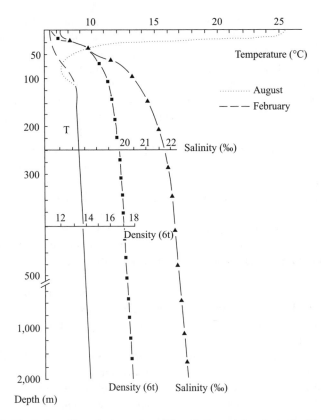

Figure 11.3 Vertical profiles of temperature (T), salinity and density in the Black Sea
Source: Modified from Balkas *et al.* (1990)

constant throughout the year (Figure 11.3). Typically, the temperature at a depth
of 1,000 m is about 9°C, and shows only a slight increase of 0.1°C per 1,000 m
towards deeper sections (Balkas *et al.*, 1990).

Marine biodiversity – General background

The Black Sea's biodiversity clearly reflects its geological history. Brackish-water
fauna known as Pontian or Caspian relics originated from the Neoeuxinian Lake,
and the components of these fauna are found only in waters with low salinity.
These fauna include bivalves such as *Dreissena*, the polychaetes (*Hypania*),
crustaceans (*Pontogammarus*), many species of goby (*Gobiidae*), sturgeons
(*Acipenseridae*), and herrings (*Clupeidae*).

Another group of species is called "cold water relics", and includes ctenophore (*Pleurobrachia*), copepods (*Calanus*), spiny dogfish (*Squalos acanthias*),

sprat (*Sprattus sprattus*), flounder (*Plactichthys flesus*), whiting (*Merlangius merlangus euxinus*), and the Black Sea trout (*Salmo trutta labrax*). These species are the second oldest inhabitants of the Black Sea, entering the sea sometime between the Neoeuxinian Lake period and the early stage of the Bosphorus formation.

As the Bosphorus established a connection with the Mediterranean Sea about 7,000–10,000 years ago, the salinity of the Black Sea rose gradually and soon the Mediterranean species were established in the Black Sea. Today, 80 per cent of total fauna species in the Black Sea are of Mediterranean origin. These Mediterranean settlers include sponges, scyphozoan, molluscs, crustaceans, echinoderms, and over 80 fish species. Some pelagic fish species migrate seasonally between the Black Sea and the Aegean-Mediterranean Sea through the Turkish Straits system (Table 11.1). Some of the groups inhabiting the Marmara Sea have, however, not been adapted to the Black Sea, such as radiolarians, pelagic foraminiferans, pteropods, siphonophores, cephalopods, and corals.

The last group of species is called "exotic or alien species", which includes species introduced either intentionally or accidentally by humans. These species are discussed later.

A total of 3,774 species have been identified, including 1,710 fungi, algae, and higher plants, 1,985 species of invertebrates, 170 fish species, and four mammals. Besides these, there are some taxa, mostly bacteria and protozoan groups, which are not yet clearly known.

Compared to the Mediterranean and Marmara Seas, the biodiversity of the Black Sea is low (Table 11.2). The reason for this low diversity is clearly related to the high level of H_2S, which prohibits many benthic species from settling, as well as to low temperatures during winter and to low salinity.

There are some key species forming faunal and floral assemblages (biocenosis) that are considered important for the sea's biodiversity. Some of these are mussels (*Mytilus*), red algae (*Phyllophora*), and brown algae (*Cystoseira*).

In addition, some wetlands, such as the Danube delta, are considered to be very important for many seabirds and other animals, serving as their feeding and breeding grounds (Wilson and Moser, 1994).

Table 11.1 Migration of fish through the Turkish Straits system

Fish species	Migration period	
	To Black Sea	From Black Sea
Pomatomous saltator	March–May	August–October
Sarda sarda	May–July	August–November
Scomber japonicus	June–August	October–November
Thunnus thynnus	April	September
Xiphias gladius	June–July	Summer

Source: Modified from Ozturk and Ozturk (1996)

Table 11.2 A comparison between the number of species of some major taxa in the Mediterranean and Black seas

Taxa	Black Sea	Mediterranean Sea
Porifera	26	593*
Polychaeta	192	400
Mollusca	190	807*
Bryzoa	16	200
Echinodermata	12	143*
Pisces	114	447
Mammalia	4	13**

Source: The figures are after Zaitsev and Mamamev (1997) except those with *, which are after Fredj, Bellan-santini, and Meinardi (1992) and ** after the authors of the present study

Threats to the Black Sea biodiversity

Fishing activity

The Black Sea is known as a very fertile sea, and has traditionally been a rich fishing ground. Figure 11.4 shows that the fish catch increased as the effort increased, except in the late 1980s when the outbreak of an alien ctenophore occurred (referred to in greater detail later). The overall catch has, however, decreased in recent years. This is due to ecological changes and insufficient fishing management, which resulted in the depletion of some commercial fishes (Prodanov et al., 1997). Among these fishes, sprat, whiting, anchovy (*Engraulis encrasicholus*), horse mackerel (*Trachurus trachurus*), turbot (*Psetta maxima*), spiny dogfish, and red mullet (*Mullus surmuletus*) have been heavily exploited. Some of the pelagic fish stocks, such as bonito (*Sarda sarda*), blue fish (*Pomatomus saltator*), mackerel (*Scomber scombrus* and *S. japonicus*), tuna (*Thunnus thynnus*), and swordfish (*Xiphias gladius*), have been depleted throughout the Mediterranean, Aegean, and Marmara Seas, and their immigration into the Black Sea was drastically reduced. In addition, sturgeons are endangered due to both overfishing and the deterioration of the environmental conditions of their native rivers, spawning grounds, and benthic zones. Among the 26 commercially important fish during the 1970s, only six remain exploitable today (Zaitsev and Mamaev, 1997).

Besides fish, red algae (*Phyllophora*), brown algae (*Cystoseira barbata*), the *Rapana* snail, the *Mya* clam, the *Mytilus* mussel, and the *Venus* clam stocks are declining in many areas due to overexploitation.

Bottom trawling is another destructive factor endangering the benthic fauna in the coastal areas, as it causes siltation and habitat degradation for the benthic communities.

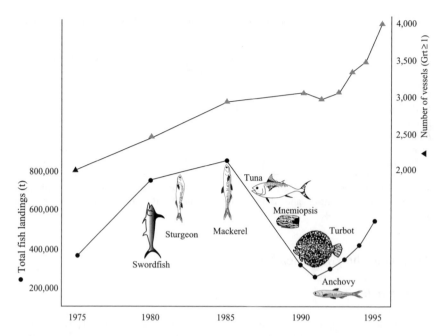

Figure 11.4 Fishing effort in terms of the number of vessels (gross tonnes >1 tonne) and total fish landing in the Black Sea

Note: The fish figures indicate the critical period of particular species, except *Mnemiopsis*, an alien jellyfish, which destabilized the Black Sea ecosystem

Source: Modified from GEF/BSEP (1996)

Eutrophication

The Black Sea is presently a hypereutrophic sea, particularly in the north-western shelf area. Mee (1992) notes that the Danube continuously discharges 60,000 tonnes of total phosphorus and some 340,000 tonnes of total inorganic nitrogen per year, contributing more than 50 per cent of the total input (Figure 11.5). There has also been an enormous increase in the nutrient load to the Black Sea in the past 25 years, probably as a result of the widespread use of phosphate detergent and agricultural intensification. The increased nutrient supply results in plankton blooms that occur especially in the shallow north-western continental shelf. Blooms are no longer special events in the Black Sea. They have become a rather regular phenomenon.

Another effect of eutrophication is hypoxia. Hypoxia eliminates mainly phytobenthic and zoobenthic communities. In 1989, due to hypoxia, large quantities of fish died near the Bulgarian and Turkish Black Sea coasts.

Zernov's *Phyllophora* field in the north-western shelf of the Black Sea has traditionally been rich in biodiversity as *Phyllophora* (red algae) is the core species

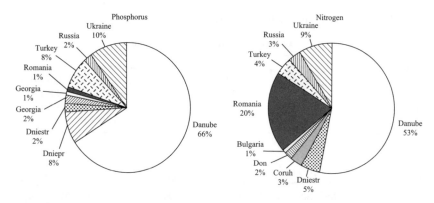

Figure 11.5 Phosphorus and nitrogen discharge into the Black Sea from the riparian countries and rivers

Source: Modified from GEF/BSEP (1996)

for *Phyllophora* biocenosis, including a wide variety of small organisms. It has also been commercially important as agar is extracted from *Phyllophora*. Due to eutrophication, however, there remained only 500 km² of Zernov's *Phyllophora* field in the early 1990s, a dramatic decrease from 11,000 km² in the 1950s (Zaitsev, 1992; cited in Zaitsev and Mamaev, 1997).

Cystoseira barbata, a kind of brown algae, is also disappearing from the Black Sea coast due to eutrophication. Like the *Phyllophora* species, this species is a key species of biocenosis, known as *Cystoseira* biocenosis. It is no longer seen on the Ukrainian coast, except the Crimea or the Romanian coast, and has become very rare in Bulgaria (Zaitsev and Mamaev, 1997).

Other types of pollution

Oil pollution has not been well documented. Some coasts, including the Marmara Sea, are heavily contaminated due to the heavy traffic of 50,000 ships annually travelling in the Black Sea (Table 11.3). The majority of the oil originates, however, from land-based sources, as well as from rivers like the Danube, which accounts for 48 per cent of the 111,000 tonnes of oil entering the Black Sea each year (Zaitsev and Mamaev, 1997). All tankers pass through the narrow Istanbul Strait, so that oil contamination is assumed to be considerable in the pre-bosphoric area. As exploitation of Caspian oil has begun – only to intensify in the twenty-first century – there will be more tankers crossing the Black Sea which may result in more serious oil pollution, including oil spills due to tanker accidents.

Metal pollution is at a considerable level in the Black Sea. For example, a study of metal content in algae in the Turkish Black Sea from 1979 to 1988 showed that

Table 11.3 Oil contamination in the Black Sea

Locality	Oil contamination (μg/l)	Source
Marmara Sea	32.01–986.53	Guven *et al.* (1997)
Novorossiik-Gelendzhik	<130	Komarov and Shimkus (1992)
Sevastopol	<540	Polikarpov *et al.* (1994)
Yalta	<180	Polikarpov *et al.* (1994)
Varna	200–300	Bojkova (1994)

metal concentrations increased during that period (Guven *et al.*, 1992). Egorov *et al.* (1994) studied the concentration of the most toxic heavy metal, mercury (Hg). The highest Hg concentrations, in the mouths of the Dniepr, Dniestr, and Danube Rivers, and near the Istanbul Strait, exceed the maximum permissible level (100 ng/dm^3). The Hg content in the Black Sea as a whole was estimated at 14,000 tonnes, and in the most biologically active layer (0–50 m) at 840 tonnes. The Hg content in the bottom sediment ranges from 20 to 290 ng/g.

Organotin compounds, such as tributyltin used in anti-fouling paints for ships and fishing nets, have been a worldwide concern lately. Madhusree *et al.* (1997) determined the concentration of butyltin compounds in harbour porpoises of the Turkish Black Sea coast. They report that the Black Sea is also contaminated by butyltin compounds, but not as much as the coastal waters of developed countries.

A study on radioactive pollution carried out on the algae collected from the coasts of the Turkish Black Sea during 1984–1989 showed that the highest accumulation of ^{137}Cs was found in *Chaetomorpha linum,* with 0.034 Bq/g, after the Chernobyl accident of 1986. The radioactivity level reached a peak soon after the accident in 1986, then gradually decreased in the following years (Guven *et al.*, 1992). Radioactivity was also detected in fish samples caught in the Black Sea after the Chernobyl accident. The total radioactivity level in the fish gradually diminished during the first three months and subsequently no Chernobyl radionuclides were detected in the fish except ^{137}Cs (Topcuoglu *et al.*, 1989). Egorov *et al.* (1994) point out that the concentration of short-lived radionuclides such as ^{95}Zr, ^{95}Nb, ^{131}I, ^{140}La, and ^{140}Ba reached a value of 10–100 Bq/kg wet weight in algae after the Chernobyl accident. Thus, generally, these studies show that the level of radioactivity was at its peak soon after the Chernobyl accident, then decreased gradually to an almost non-detectable level.

Organochlorine compounds have been used in the Black Sea basin, and enter into the sea through rivers, industrial waste, and the atmosphere. Akman *et al.* (1978) detected organochlorine insecticide in various fish species and mussel samples from the Turkish Black Sea coasts, and the ratio of incidence of the residue type in all samples were calculated as DDT and BHC derivatives 100 per cent, dieldrin 94.6 per cent, and aldrin and endrin 51.5 per cent. The main concentrations of total insecticide residue found in fish were 0.184–0.713 ppm wet weight. The north-western part of the Black Sea is most contaminated, specifically the bottom sediments (Egorov *et al.*, 1994). Tanabe *et al.* (1997a) determined a wide range of organochlorine residues in the blubber of harbour porpoises and some fish from the Black Sea. They reported that amongst all organochlorine residues, the concentrations of DDTs (8.3–180 mg/g wet weight) were the highest in harbour porpoises, and that the contamination of the Black Sea by DDTs and HCHs was relatively high in comparison to other polluted regions of the world. Tanabe *et al.* (1997b) also determined the highly toxic coplanar PCB and other isomers in the harbour porpoises and their prey fish collected from the Turkish Black Sea coast. PCB concentrations were found to range from 5.0 to 34 mg/g wet weight in the blubber of the porpoises.

These pollutants enter into the Black Sea through rivers, direct discharge of industrial and domestic wastes from both the land and ships, the atmosphere, and dumping. Topcuoglu *et al.* (1990) noted that they detected heavy metals such as Pb and Cd in barrels dumped in the western Black Sea containing organic pollutants like DDT and PCB.

Alien species

As many ships continuously travel through the Black Sea, there is a good chance that an alien or exotic species will be introduced to the Black Sea. The first documented case of this is typical: a sea snail, *Rapana thomasiana*, was found in the Black Sea at the end of the 1960s. It is assumed that this sea snail was introduced by attaching itself to a ship's hull in the Sea of Japan. This species feeds on mussels, oysters, and other shells, thus destroying the beds of these shells. After the 1970s, this species reached as far as the Marmara Sea. Today – of great economic value and greatly exploited – it is cultivated in the Black Sea and sold to the eastern Asian market.

A relatively recent immigrant to the Black Sea is a ctenophore, *Mnemiopsis leidyi*, assumed to enter the Black Sea with tanker ballast water from the North Atlantic Ocean. At its peak, this predatory ctenophore increased up to 1 kg/m^2 in the open sea and 5 kg/m^2 in shelf areas (Vinogradov *et al.*, 1989). This species altered the balance of the trophic web by feeding on fish eggs and larvae, which resulted in the collapse of pelagic fish stocks in the late 1980s. This species also penetrated into the Azov Sea in 1988 and exterminated the zooplankton stocks there (Volovik, Volovik, and Myrzoyan, 1995). It also reached the Marmara Sea in 1991 and destabilized its pelagic ecosystem (Ozturk, 1992; Shiganova *et al.*, 1995). *Mnemiopsis* was observed in bulk quantities even in the eastern Mediterranean around Mersin Bay in 1992 (Kideys and Niermann, 1993). The population of *Mnemiopsis* has now declined in the Black Sea and the adjacent waters.

There are some other opportunistic settlers found in the Black Sea, such as clams (*Cuniarca curnea* and *Mya arenaria*) and crabs (*Rhithropanopeus harrisi* and *Callinectes sapidus*), as well as those introduced intentionally by humans, such as the haarder, *Mugil soiuy*. So far, there have been no reports of negative effects from these new species on the local ecosystem. Still, controlling the introduction of exotic species is a vital problem in the Black Sea. For this reason, precautions against the discharge of tanker ballast water should be adopted according to International Maritime Organization (IMO) guidelines.

Coastal degradation and erosion

Coastal degradation is an important factor for the Black Sea's marine biodiversity. In recent years, due to the intense development of tourism in the Black Sea area, which results in over-urbanization and road construction along the seashores, habitat loss began for several plants and animals.

Coastal degradation also causes coastal erosion and landslides in some areas, such as on the Turkish coasts. Another cause of coastal degradation is the discarding of solid wastes. These solid materials are washed ashore, eventually ruining the amenity of beaches.

Threats to marine mammals

Marine mammals, as the top predators of the ecosystem, are very sensitive to ecological change. Thus they represent the ecosystem and its problems, which is why they are reviewed here in particular.

There are three cetacean species and one pinniped living in the Black Sea. These cetaceans are bottlenose dolphins (*Tursiops truncatus*), common dolphins (*Delphinus delphis*), and harbour porpoises (*Phocoena phocoena*). Cetaceans of the Black Sea have been affected by water pollution, food shortages, microbial pollution, loss of habitats, and incidental killing. As the only pinniped species, the Mediterranean monk seal (*Monachus monachus*) survives in the Black Sea.

Commercial dolphin fishing operated along the Black Sea coasts until a ban was imposed in 1966 in the ex-Soviet Union countries, Romania and Bulgaria, and in 1983 in Turkey. The catching method was purse seining and, later, shooting by rifle. The estimate of the catch varies depending on the source: for example, Ivanov and Beverton (1985) estimate the annual catch for 1938 at 7,200 tonnes, and during 1952–1963 at 1,342 tonnes for all Black Sea countries, while Berkes (1977) estimates 59,000 tonnes in 1966 and 44,000 tonnes in 1973 for Turkey alone.

Since the ban of the commercial catch, the Black Sea dolphins suffer mainly from incidental catch, especially off the Turkish coasts, although there are no data systematically collected. Every year, at least several hundred dolphins, mainly harbour porpoises and, more rarely, bottlenose dolphins, drown in the bottom gill nets set for sole, turbot, and sturgeon between early April and June off the Turkish coast, according to the authors' survey (unpublished).

Cetaceans depend on sound for communication and feeding. The heavy boat traffic also influences the movement of dolphin schools. This problem is very obvious in the Istanbul Strait, as there are about 50,000 ships travelling between the Marmara and Black Seas through the strait every year. This heavy sea traffic hinders the movement of dolphin populations between the Marmara and Black Seas (Ozturk and Ozturk, 1996).

Poor water quality due to domestic wastewater causes microbial pollution. Sixty species of microorganisms, including various intestinal bacteria, were detected in the respiratory organs of bottlenose dolphins and harbour porpoises (Birkun *et al.*, 1992).

Another marine mammal species in the Black Sea, the Mediterranean monk seal, is on the verge of worldwide extinction. Distribution of the Mediterranean monk seal in the Black Sea had been studied by some researchers, for example, Caspers (1950), Berkes *et al.* (1978), Israel (1992), and Ozturk (1992, 1996). Its most important habitats were Cape Caliacra in Bulgaria, the Danube estuary in Romania, the Tartankuta Peninsula in Ukraine, and the Turkish coast of the western Black Sea up to Sinop (Ozturk, 1992). Today only a few specimens remain in the Turkish Black Sea coasts, on small isolated beaches with cliffs between Zonguldak and Ayancik.

Decline of the Mediterranean monk seal is related to the loss of habitats, their capture for zoos before protection of the species was legally enforced in 1978 in Turkey, isolation from the Mediterranean population resulting in a genetic bottleneck, and insufficient protection enforcement.

Future of the Black Sea

As seen above, the problems that the Black Sea is facing due to various anthropogenic influences are complex. There is no single remedy that will solve all these problems. The appropriate management of fisheries and pollution assessment, however, and international cooperation, can contribute – at the very least – to conserving the present status of the Black Sea. For this purpose, a number of conventions have been proposed and signed by the riparian countries. Among them, the Bucharest Convention, the Agreement on the Conservation of the Cetaceans of the Black Sea, Mediterranean, and Contiguous Atlantic Sea (ACCOBAMS), and the Black Sea Strategic Action Plan have a vital role in the protection of the biodiversity of the Black Sea.

The Bucharest Convention on the Protection of the Black Sea Against Pollution was adopted in 1992 and is a broad convention intended to conserve the marine environment and the living resources of the Black Sea through protection against pollution. It was followed by the Odessa Ministerial Declaration of 1993, and, as the first step to its implementation, the ministers of environment of the Black Sea countries signed the Strategic Action Plan for Rehabilitation and Protection of the Black Sea in 1996.

ACCOBAMS is specifically designed for cetaceans, as they are recognized as an integral part of the marine ecosystem, and must be protected due to their vulnerability. It was signed in 1996 by 16 countries, including three Black Sea riparian countries, Georgia, Romania, and Ukraine.

It must be stressed that these international agreements and strategies have to be effectively implemented at the earliest date, which, due to the socio-economic situations in the Black Sea countries, is often very difficult. Besides these, international cooperation for scientific research and monitoring studies is urgently needed.

As far as fishing is concerned, there is an initiative for a fisheries convention that will lead to more effective management of Black Sea resources. As fishing is one of the few financial sources for the Black Sea peoples, sustainable development must be considered. Nevertheless, some restrictions should be enforced during certain seasons for certain areas, species, and sizes. For example, the sturgeon and Black Sea trout fisheries have been forbidden in Turkey since 1997.

Protection of the Turkish Straits system also has a vital role in the protection of marine biodiversity in the Black Sea, as it is the only source of water renewal, and as it serves as a biological corridor for many migratory species (Ozturk and Ozturk, 1996).

Another critical factor for the protection of the Black Sea's biodiversity is pollution control in the Danube. As we have seen, this international river contributes large amounts of pollutants, both organic and inorganic. Unless the pollution input by the Danube is controlled, the condition of the Black Sea cannot be improved.

Specially protected areas should be established for highly vulnerable species, such as the Mediterranean monk seal, dolphins, and sturgeons. Fishing and recreational activities should be restricted in these areas and regular monitoring studies executed.

The future of the Black Sea does not just depend on one nation's effort – or on a single agreement. Rather, it depends on international cooperation and intergovernmental coordination on a global scale.

ACKNOWLEDGEMENTS

The authors thank the Turkish Marine Research Foundation for their support.

REFERENCES

Akman, S., S. Ceylan, Y. Sanli, S. Gurtunca, and F. Aksiray. 1978. Karadenizde avlanan baliklarda ve bu baliklardan elde edilen balik yagi ve balik unlarinda klorlu hidrokarbon insektisid rezidulerinin arastirilmasi. (Determination of the organochlorine insecticide residues in the various fish species obtained from the Black Sea and in the fish oil and meal). TUBITAK Pub. No. 401, 28pp. (in Turkish).

Balkas, T., G. Decev, R. Mihnea, R. Serbanescu, and U. Unlata. 1990. State of the *Marine Environment in the Black Sea Region*. UNEP Regional Seas Reports and Studies, No. 124, 40pp. UNEP, Nairobi.

Berkes, F. 1977. Turkish dolphin fisheries. *Journal of the Fauna Preservation Society*. London, 13(2): pp. 163–167.

Berkes, F., H. Anat, H. Esenel, and M. Kislalioglu. 1978. Distribution and ecology of *Monachus monachus* on Turkish coasts. In: *The Mediterranean Monk Seal*. K. Ronald, R. Duguy (eds.), pp. 113–127. Pergamon Press, England.

Birkun, A.Jr, V. Krivokhizin, A.B. Shvatsky, N.A. Miloserdova, Y. Radygin, V. Pavlov, N. Nikitina, B. Goldin, M. Artov, Y. Suremkina, Y. Zhivkova, and S. Plebanksy. 1992. Present status and future of Black Sea dolphins. In: *Proceedings of the Sixth Annual Conference of the European Cetacean Society, San Remo, Italy,* P.G. Evans (ed.), pp. 47–52.

Bojkova, D. 1994. Oil products pollution of sea waters in Varna Bay. In: *Proceedings of the Black Sea Symposium*, K.C. Guven (ed.), pp. 137–145. The Black Sea Foundation, Istanbul.

Caspers, H. 1950. Beobachtungen uber das vorkomen der Monchrobbe/*Monachus albiventer*/ im Schwerzen Mer. In: *Neue Ergebnis und Problemme der Zoologie*, pp. 91–105.

Egorov, V.N., G.G. Polikarpov, L.G. Kulebakina, S.K. Svetasheva, and M.V. Zherko. 1994. Contamination of the Black Sea with radionuclides, mercury and polychlorbiphenils and the role of biotic factor in the self-decontamination of waters. In: *Proceedings of the Black Sea Symposium*, K.C. Guven (ed.), pp. 43–58. The Black Sea Foundation, Istanbul.

Erinc, S. 1984. Karadeniz canaginin Jeomorfolojik ve yapisal ozellikleri ve morfometrisi. (Geomorphological and structural characteristics of the Black Sea Basin and its morphometry.) *I.U. Cografya Enstitusu Dergisi*, 1: 15–22 (in Turkish).

Fredj, G., D. Bellan-Santini, and M. Meinardi. 1992. Etat des connaisances sur la faune marine mediterranee. *Bulletin de l'Institut oceanographique*, Monaco, Special Issue 9, pp. 133–145.

GEF/BSEP (Global Environmental Facility/Black Sea Environmental Programme). 1996. Black Sea Transboundary Diagnostic Analysis. *GEF/BSEP Programme Coordination Unit*, Istanbul, 104pp.

Guven, K.C., E. Cevher, S. Yurdoglu, N. Gungor, N. Kose, M. Bulut, N. Bayulgen, S. Topcuoglu, B. Guvener, B. Ozturk, I. Bildaci, and H. Evliya. 1992. The radioactivity level in marine algae on the Turkish coasts during 1981–1989. *Bulletin of the Institute of Marine Sciences and Geography*, Istanbul University, 9(9): 1–10.

Israel, L. 1992. Thirty years of Mediterranean monk seal protection: A review. Netherlands Commission for Nature Protection, No. 28, 65pp.

Ivanov, L. and R.J.H. Beverton. 1985. *The Fisheries Resources of the Mediterranean, Part 2: The Black Sea. Studies and Reviews*, No. 60, 135pp. General Fisheries Council for the Mediterranean, FAO, Rome.

Izdar, E. and M. Ergun. 1991. Recent geological evolution of the Black Sea: An overview. In: *Black Sea Oceanography*, E. Izdar and J.W. Murray (eds), pp. 379–387. Kluwer Academic Publishers.

Kideys, A.E. and U. Niermann. 1993. Intrusion of *Mnemiopsis macradyi* (Ctenophora; Lobata) into the Mediterranean Sea. *Senckenbergiana Maritima*, 23(1/3): 43–47.

Komarov, A.V., and K.M. Shimkus. 1992. Features of seasonal input and accumulation of pollutants in the Novorossiysk–Gelendzhik region of the Black Sea and their ecological consequences. In: *Abstracts of Assessment of Land-Based Sources of Marine Pollution in the Sea Adjacent to the CIS*, Sevastopol, 6–10 April 1992, 1: 40–41. (in Russian).

Madhusree, B., S. Tanabe, A.A. Ozturk, R. Tatsukawa, N. Miyazaki, E. Ozdamar, O. Aral, O. Samsun, and B. Ozturk. 1997. Contamination by butyltin compounds in harbour porpoise (*Phocoena phocoena*) from the Black Sea. *Fresenius Journal of Analytical Chemistry*, 359: 244–248.

Mee, L. 1992. The Black Sea in crises: a need for concerted international action. *AMBIO*, 21(4): 278–286.

Oguz, T., P. Malanotte-Rizzoli, and D.G. Aubrey. 1994. Wind and thermohaline circulation of the Black Sea driven by yearly mean climatological forcing. *Journal of Geophysical Research*, 100(C4): 6845–6863.

Ozturk, B. 1992. *Akdeniz Foku (Mediterranean Monk Seal)*, 215pp. Anahtar Yayinlari, Istanbul (in Turkish).

Ozturk, B. 1996. Past, present and future of the Mediterranean Monk Seal *Monachus monachus* (Hermann, 1779) in the Black Sea. In: *Proceedings of the First International Symposium on the Marine Mammals of the Black Sea*, B. Ozturk (ed.), pp. 96–102. UNEP, GEF, IU.

Ozturk, B. and A.A. Ozturk. 1996. On the biology of the Turkish straits system. *Bulletin de l'Institut oceanographique*, Monaco, Special Issue 17, CIESM Science Series No. 2: pp. 205–221.

Polikarpov, G.G., Yu.P. Zaitsev, V.I. Zats, L.A. Radchenko. 1994. Pollution of the Black Sea (levels and sources). In: *Proceedings of the Black Sea Symposium*, K.C. Guven (ed.), pp. 15–42. Black Sea Foundation, Istanbul.

Prodanov, K., K. Mikhailov, G. Daskalov, C. Maxim, A. Chaschin, A. Arkhippov, V. Shlyakhov, and E. Ozdamar. 1997. *Environmental Management of Fish Resources in the Black Sea and Their Rational Exploitation*. Studies and Reviews No. 68, 178pp. General Fisheries Council for the Mediterranean, FAO, Rome.

Shiganova, T., A.N. Tarkan, A. Dede, and M. Cebeci. 1995. Distribution of the ichthyo-jellyplankton *Mnemiopsis leidyi* (Agassiz, 1865) in the Marmara Sea (October 1992). *Turkish Jounal of Marine Sciences*, 1(1): 3–12.

Soljankin, E.V. 1963. On the water balance of the Black Sea. *Okeanologiya*, No. 3: pp. 986–993 (in Russian).

Sorokin, Y. 1983. The Black Sea. In: *Ecosystems of the World 26: Estuaries and Enclosed Seas*, B.H. Ketchum (ed.), pp. 253–292. Elsevier Science, Amsterdam.

Tanabe, S., B. Madhusree, A.A. Ozturk, R. Tatsukawa, N. Miyazaki, E. Ozdamar, O. Aral, O. Samsun, and B. Ozturk. 1997a. Persistent organochlorine residues in harbour porpoise (*Phocoena phocoena*) from the Black Sea. *Marine Pollution Bulletin*, 34(5): 338–347.

Tanabe, S., B. Madhusree, A.A. Ozturk, R. Tatsukawa, N. Miyazaki, E. Ozdamar, O. Aral, O. Samsun, and B. Ozturk. 1997b. Isomer-specific analysis of polychlorinated biphenyls in harbour porpoise (*Phocoena phocoena*) from the Black Sea. *Marine Pollution Bulletin*, 34(9): 712–720.

Topcuoglu, S., M.Y. Unlu, N. Sezginer, M. Sonmez, A.M. Bulut, N. Bayulgen, R. Kucukcezar, and N. Kose. 1989. Karadeniz, Bogazci ve Marmara deniz urunlerinde

Cernobil sonrasi yapilan radyoaktivite olcumleri (radioactivity measurement in the sea products from the Bosphorus, Marmara and Black Seas after the Chernobyl accident) In: *Abstracts of the Third National Nuclear Sciences Congress,* Istanbul, pp. 751–756. (in Turkish).

Topcuoglu, S., N. Saigi, N. Erenturk, and A.M. Bulut. 1990. Karadenizíe atlin varillerle ilgili toksisite calismasi (Study on the toxicity of barrels dumped into the Black Sea). *Cekmece Nuclear Research and Training Center, Report* No. 276, 5pp. (in Turkish).

Unlata, U., T. Oguz, M. A. Latif, and E. Ozsoy. 1990. On the physical oceanography of the Turkish Straits. In: *Physical Oceanography of the Straits*, J.L. Pratt, NATO/ASI Series (ed.), pp. 25–60. Kluwer Academic Publishers.

Vinogradov, M.E., E.A. Shuskina, E.I. Musaeva, and Y.I. Sorokin. 1989. Ctenophore *Mnemiopsis leidyi* (Agassiz) (Ctenophora, Lobata) – New settlers in the Black Sea. *Oceanology*, 29(20): 293–299.

Volovik, Y.S., S.P. Volovik, and Z.A. Myrzoyan. 1995. Modelling of the *Mnemiopsis* sp. population in the Azov Sea. *ICES Journal of Marine Science*, 52: 735–746.

Wilson, M. and E. Moser. 1994. *Conservation of Black Sea Wetlands*. IWRB Publication No. 33, 76pp. International Waterfowl and Wetlands Research Bureau, Slimbridge, U.K.

Zaitsev, Y. 1992. Impacts of eutrophication on the Black Sea fauna. In: *Fisheries and Environmental Studies in the Black Sea System*. Studies and Reviews No. 64, pp. 59–86. General Fisheries Council for the Mediterranean, FAO, Rome.

Zaitsev, Y. and V. Mamamev. 1997. *Marine Biological Diversity in the Black Sea*. United Nations Publication, New York. 208pp.

12

Marine biodiversity of the south-western Atlantic Ocean and main environmental problems of the region

Ricardo Bastida, Diego Rodrìguez, Norbeto Scarlato, and Marco Favero

Introduction

In the past, and even in the present, the different environments of our planet have been studied independently, as if there were no links and dependencies between them. In contrast we now know that South American oceanographic phenomena, such as El Niño, affect the rest of the world. Nowadays, we also understand that actions produced by man during the "gold rush" in isolated inland areas of Brazil have had repercussions in distant marine communities of the south-western Atlantic Ocean (SWAO).

Global atmospheric change and the ozone layer problems are subjects of concern for scientists and are no longer matters of debate for the rest of the community. These and other subjects have served to show that environmental problems occur much more rapidly than the solutions of scientists and politicians.

In this context, biodiversity has been revalorized and is becoming a term of frequent use. Nowadays, the term is being used by scientists, governmental authorities, and non-governmental organizations (NGOs). In spite of this, marine biodiversity has not worried our community in the same way as rainforests, perhaps because of the mistaken belief in the ocean's relative immunity to human activity. Nothing, however, is further from the truth, since oceans and seas are nothing but the sinks of our little planet and hold highly diversified ecosystems that are very sensitive to environmental changes.

Perhaps one of the aspects that differentiate the marine environment is that there is no clear fragmentation, rather, it is a very wide environment, interconnected and without barriers. Any impact, sooner or later, affects the rest of this complex ecosystem. When we think about that, a simple question comes to mind: how much do we know about the oceans? It is doubtless more than we knew at the beginning of the twentieth century, but much less than what we should know at the end of it. Probably the old assumption that we know more of outer space than of inner space is still valid. Mankind would rather know if there is life on Mars than know about the undiscovered species that live on our own planet.

Another question we could ask ourselves is whether biodiversity has historically worried mankind. The answer is yes, since this is one of the aspects of the human condition that concerned naturalists such as Cuvier, Linné, d'Orbigny, Darwin, and Humboldt, among others. Unfortunately, it is man himself who has not given enough time to characterizing the animals with whom we share this delicate planet. Many environmental crises have taken place in recent decades, and this has forced international organizations to get involved in the subject. Following this trend, the United Nations Convention on Biological Diversity, held in 1994, defined biological diversity in the following terms:

The variability among living organisms from all sources including, *inter alia*, terrestrial, marine and other aquatic ecosystems and the ecological complexes of which they are part; this includes diversity within species, between species, and of ecosystems.

This chapter refers now to the biodiversity and environmental problems affecting a vast oceanic region of the southern hemisphere: the south-western Atlantic Ocean (SWAO).

An updated overview of the oceanographic characteristics of the SWAO will be included in this chapter, since they condition the main biogeographic areas of the region, which in turn show particular characteristics in their biodiversity patterns. Finally, the main environmental impact factors identified in the region and their incidence in biodiversity and environmental quality are analysed.

Brief history of SWAO exploration and research

The SWAO, constituted by the coastal and shelf areas of Argentina, Uruguay, and southern Brazil, has attracted European expeditions since the seventeenth century. At those times, the only route of communication between the Atlantic and the Pacific was via the Strait of Magellan, and more recently the dangerous Cape Horn. Detailed descriptions of the seafloor and coastlines were needed, and these European expeditions made nautical charts and exploited some of the renewable resources of the region.

Among these, marine mammals and birds were especially attractive and thus were intensively exploited from the eighteenth century on, bringing many of them to the verge of extinction. This activity not only affected animal populations, but also harmed the aboriginal populations, such as the Fuegian, who relied on the local pinniped rockeries for food. As a consequence of seal overexploitation, the natives were forced to turn to a low-calorie diet based mainly on coastal molluscs. On the other hand, Europeans tried to assimilate these coastal aborigines with the aim of avoiding their attacks. They also tried to train them to help sailors during the frequent shipwrecks occurring in the stormy austral waters.

Scientific expeditions to this sector of the Atlantic started during the second half of the eighteenth century and increased remarkably during the nineteenth century. French, English, German, and American expeditions periodically surveyed these seas. Among them, the famous expeditions of HMS *Beagle* (1831–1836) and of HMS *Challenge* (1876) signalled the onset of modern oceanographic research. A more detailed knowledge of the SWAO and the first flora and fauna collections derived from these and other expeditions of the nineteenth century.

In the beginning of the twentieth century there was an increase in scientific expeditions, due partly to the increasing surveys of Antarctica via South America. Also at that time, the first hydrographical studies were performed in Argentina, Uruguay, and Brazil. Undoubtedly the *Discovery* expedition (1931–1935) was very important for the development of marine sciences in the SWAO. From this famous expedition on, and until the 1950s, there was little advance in the general biological knowledge of this region and of its fishery resources.

Since the 1960s, however, a clear advance in the development of the marine sciences in several South American countries can be observed. UNESCO promoted the creation of several research groups, which increased the oceanographical studies in the area. The FAO's Fisheries Development Project (1966–1975) focused on the SWAO's renewable resources. During this period, European ships such as the *Calypso* (1960–1961), *Walther Herwig* (1966 and 1970–1971), and *Prof. Siedlecki* (1973) updated the knowledge of biodiversity in the SWAO and assessed its fishery resources. Simultaneously, Japanese research ships such as the *Orient Maru* (1976–1977) and the *Shinkai Maru* (1976–1977) made important contributions to this subject in cruises where research took place in Argentina, and from where many of the data of the present work derive.

Basic oceanographic characteristics of the SWAO

The SWAO is regulated by different water masses. The Malvinas-Falklands Current (MFC) is a northward-running branch of the subantarctic Cabo de Hornos Current, which influences both coastal and offshore areas (Figure 12.1). As it moves northward, the MFC separates from the coast and connects with offshore waters, reaching Cabo Frío (Brazil, 23°S, 42°W) as a coastal upwelling. This flowing pattern has seasonal variations, with annual mean temperature ranges from 4°C to 11°C and salinity ranging yearly from 33.8 to 34.4 ‰

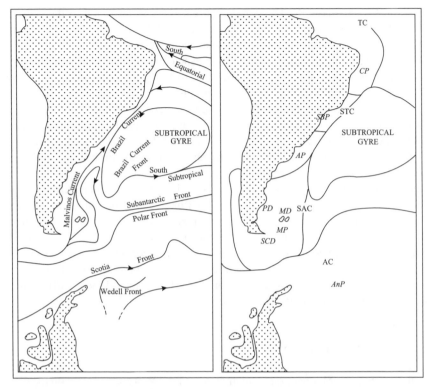

Figure 12.1 Main oceanic currents in the SWAO, and biogeography of the SWAO

Notes: AC = Antarctic Complex; SAC = Subantarctic Complex; STC = Subtropical Complex; TC = Tropical Complex; MP = Magellanic Province; ANP = Antarctic Province; AP = Argentine Province; SBP = South Brazilian Province; CP = Caribbean Province; SCD = South Chilean District; PD = Patagonian District; MD = Malvinas-Falklands District

Source: After Peterson and Stramma (1991)

(Bastida, Roux, and Martinez, 1992; Boltovskoy, 1981). The high primary productivity of the MFC supports very important fisheries in Patagonia.

The Brazil Current (BC) is a southbound branch of the South Equatorial Current and moves along the Brazilian coast (Figure 12.1). The BC and MFC meet at the Subtropical Convergence located approximately at 35°S. The convergence affects mainly oceanic areas and has some influence on the continental shelf, where tropical fauna are sporadically found. BC waters are less productive than those of the MFC, with mean temperature from 14°C to 25°C, and salinity from 35 to 35.5 per mill (Bastida, Roux, and Martinez, 1992; Boltovskoy, 1981).

A third water mass, the Patagonian Current (PC), is restricted to the coastal zone of Patagonia, and flows between the coast and the MFC (Figure 12.1). The PC is a relatively stable water mass of subantarctic origin that moves mostly northward and reaches up to 38°S. Its mean temperature, variable both with latitude and season, ranges annually from 5°C to 16°C, and its salinity from 33 to 33.5 per mill (Bastida, Roux, and Martinez, 1992; Boltovskoy, 1981). The major extension of these water masses flows upon a continental shelf of variable width. Its maximum width is 460 miles (at the Malvinas-Falkland Islands) and its minimum extension is about five miles, in both Tierra del Fuego and northern Brazil.

The coastal sector of the region has a highly variable geomorphology, while the continental shelf is very homogeneous. Although the southern sector presents a smooth slope, linear shoals, stairs, channels, and ridges are found at different latitudes, and probably belong to ancient coastal lines. Sands dominate the bottom sediments, but many different materials are mixed with the basal sandy sediment throughout the shelf. Some places show a high percentage of gravel, formed either by small pebbles or invertebrate bioclasts. The concentration of carbonate in the sediment, mostly of biological origin, is lower than that in tropical zones, but some areas show fairly high concentrations (Bastida and Urien, 1981). Hard bottoms are scarce and much less studied.

Biogeography of the SWAO

Initially, patterns of biogeographical division of the SWAO were based on the distribution of molluscs (Carcelles and Williamson, 1951), while other faunistic groups were later studied. The first integrated analysis of the biogeography of the region was carried out in 1962 by a special UNESCO meeting held in Mar del Plata (Argentina), where experts in algae and marine invertebrates and vertebrates took part. The basic biogeographical scheme of South America derived from this meeting, and recent ecological studies carried out on plankton, benthos, and nekton have complemented its view (Bastida, Roux, and Martinez, 1992; Boltovskoy, 1981; Menni and López, 1984).

Although the dynamics of its main water masses do regulate the biogeographical scheme of the SWAO, it must be noted that not all organisms respond biogeographically in the same way. A similar situation can be mentioned for biodiversity, as certain groups respond differently in the biogeographical subdivisions referred to in the following paragraphs. The authors will try to present a simplified scheme responding to the generalities of most of the groups inhabiting the SWAO.

Four different complexes can be identified in the SWAO (Figure 12.1). Each one of these complexes is linked with the water masses' dynamics, and may be divided into biogeographical provinces on some occasions; smaller units are called biogeographical districts.

Antarctic Complex

The Antarctic Complex ranges from Antarctica to the Antarctic Convergence or Polar Front (Figure 12.1). It is a highly productive zone, with its primary planktonic productivity limited to an euphotic layer 50 m thick and restricted to the period when the Arctic region enjoys 24 hours of sunlight (December–March) (El-Sayed, 1968; Ferguson-Wood and Walsh, 1968). Apart from its extreme climatic conditions, this area is characterized by unusually deep continental shelves (which can run to depths of 800 m) formed as a result of the isostatic depression of the continent derived from the mass of the polar ice cap (Clarke, 1996a).

This complex is the only one in the SWAO lacking typical coastal fluvial habitats, and intertidal benthic communities are very scarce as a result of the direct ice impact. Some faunistic groups are poorly diversified, but the complex boasts nevertheless a high biomass (zooplankton and most of the vertebrates). Other groups, such as benthic invertebrates, show an important diversity that is unusual for these latitudes (Clarke, 1996a, 1996b). Benthic assemblages in Antarctica show that life is both abundance-rich and taxonomically diverse, and although a fully developed hard substratum in such high latitudes may not quite rival a tropical reef in diversity, it is nevertheless rich in species (Clarke, 1996b).

Due to the uniform water mass surrounding it, the Antarctic Complex shows great similarity across all Antarctica. In spite of this, some people propose further subdividing the complex into biogeographical provinces (Dell, 1972; Hedgpeth, 1970).

Subantarctic Complex

This complex includes waters located between the Antarctic Convergence and the Subtropical Convergence (Figure 12.1). Its waters, though productive, are less rich than Antarctic waters, though its primary production values throughout the year are more homogeneous due to a longer period of maximum radiation that runs from September to February. Still, it must be noted that the South American zone is less productive than its counterpart in South Africa. Fish and plankton diversity tends to increase, and benthic invertebrates keep high diversity levels.

The Subantarctic Complex consists of two provinces: the Magellanic and Argentinean. The Magellanic Province, in its Atlantic sector, extends from Tierra del Fuego to the Península Valdés (42°S) in its coastal sector, but extends much further north offshore to 34°S due to its association with the Malvinas-Falklands Current flow (Figure 12.1). This province can be further divided into two districts: the Patagonic and the Malvinas-Falklands. The first comprises the shallower shelf sector of Argentina; the second the deep shelf and insular areas (Bastida and Urien, 1981; Bastida, Roux, and Martinez, 1992; Boltovskoy, 1964).

The Magellanic Province also extends to the Pacific coast of South America, to Puerto Montt (41°S), and is divided into two districts, the South Chilean and

Central Chilean. The Atlantic and Pacific coasts of southern South America share a high number of species, especially those of coastal origin, belonging to the algae, invertebrate, and vertebrate groups. According to Ihering's theory (Szidat, 1964), this relation between the two oceans at the southern tip of South America is quite recent, and took place in the latter part of the Pleistocene era after Tierra del Fuego separated from the Antarctic Peninsula. Thus, based on the origins of the parasitic fauna from both groups, it may be stated that the SWAO has a double origin: Pacific on one hand, North Atlantic on the other.

The Argentine Province extends from the north of Península Valdés (42°S) to 31°S in southern Brazil. This province extends throughout the coastal area and the shallow and intermediate continental shelf of Argentina, while the deep-shelf sector at this latitude corresponds to the Magellanic Province (Bastida, Roux, and Martinez, 1992; Roux *et al.*, 1988). This is an area of warmer waters, which turn it into a transition zone between the Subantarctic and the Subtropical Complexes, although its components belong to the latter. Its flora and fauna species are clearly different from those of the Magellanic province, though a very low percentage of mixture between both provinces may be noted. Some authors have proposed subdivisions into two districts (Balech, 1964), but this is not evident for many faunistic groups. Important coastal estuarine ecosystems are included in this biogeographical province (e.g. the estuary of the Rio de la Plata, Lagoa dos Patos, etc.) and it marks the southern limit of the geographical distribution of mangrove communities.

Subtropical Complex

This complex extends between the tropics, approximately between 10°S and the Subtropical Convergence, whose northern limit in offshore waters is located between 30 and 35°S (Boltovskoy, 1981; Figure 12.1). The primary production of this complex is lower than those of the previous complexes. The permanent thermocline, located 60–100 m deep, does not allow for a vertical circulation of the rich, deep waters, limiting the input of nutrients to this layer.

An increasing trend of biodiversity towards the north may be observed. In the coastal zones of this complex, mangrove communities begin to appear. The Subtropical Complex includes the South-Brazilian biogeographical province, including coastal and shelf areas from southern Brazil to Cabo Frío, near Rio de Janeiro (23°S), where the last upwelling zone of the Malvinas-Falklands Current may be observed. Some authors, it should be noted, do not recognize the South-Brazilian biogeographical province to be a province as such, but consider it merely as a district of the Argentinean biogeographical province (Menni, Ringuelet, and Arámburu, 1984). This sector results in an area with transitional or ecotonal characteristics that resemble those of the richer tropical ecosystems.

Tropical Complex

The Tropical Complex extends over the equator to include sections of both hemispheres. In the SWAO it originates in the South Equatorial Current and its southern branch, the Brazil Current (Figure 12.1). Primary production is mainly composed of dinoflagellates and cianophytes; diatoms are very low (10 per cent) in this complex, while in the Antarctic and Subantarctic Complexes they are more than 50 per cent.

Here, an increase in biodiversity is observed for all biological components. This complex includes coral reefs and mangroves with very rich invertebrate communities and fish assemblages associated with them. Throughout the Tropical Complex, including the Caribbean biogeographical province, low biomass values are compensated by high biodiversity in many of its ecosystems (Boltovskoy, 1964; Stuardo, 1964), making the Tropical Complex highly sensitive to the potentially damaging effects of human activity.

General aspects of SWAO planktonic assemblages

Both systematic and biological studies of plankton have long traditions in the SWAO, where most groups have been deeply sampled for a wide variety of studies (Boltovskoy, 1981). Without any doubt, we know more about plankton than benthos and nekton, where many taxa continue to be poorly understood.

Phytoplankton and zooplankton have a clear correlation with the dynamics of water masses. Their diversity and biomass are regulated directly, and in short-term periods, by physical and chemical changes in the environment (El-Sayed, 1968). There is also a good correspondence in these organisms with the biogeographical provinces of the SWAO, as many plankton species are very good biological indicators (Balech, 1976).

The relation between diversity and biomass can be clearly observed in plankton communities of the SWAO. As shown in Figure 12.2, biomass reaches its highest values in the Antarctic Complex and declines around 50 per cent in the Subantarctic Complex, although primary and secondary production are still very high. Biomass values of the Subtropical Complex decline another 50 per cent with areas of low productivity (Boltovskoy, 1981).

Plankton diversity, on the contrary, shows its minimum values in Antarctic waters, half of those in the Subantarctic Complex (Figure 12.2). The Subtropical and Tropical Complexes show the highest plankton diversity values, the Subtropical being the more important (about 6.5 times the Antarctic values) in relation to the ecotonal characteristics of the Subtropical Convergence (Boltovskoy, 1981). Latitudinal diversity declines can also be observed in plankton communities, in relation to several environmental gradients of the region.

Red tides have increased dramatically over the last two decades, most of them related to the dinoflagellate *Alexandryum excavatum*, frequently producing

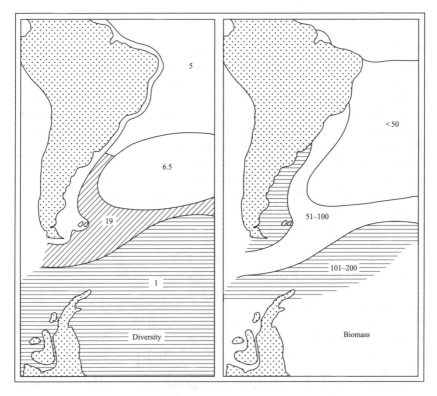

Figure 12.2 Zooplankton diversity, and zooplankton biomass in the SWAO
Source: After Boltovskoy (1981)

human sickness and deaths. The increase in such phenomena seems to be related
to pollution and the dynamics of certain chemical elements of seawater (Carreto
et al., 1985).

General aspects of SWAO benthic communities

Ecological studies of the marine benthic communities of the SWAO began in the
early 1960s with descriptions of coastal communities in Argentina, Brazil,
Venezuela, and Antarctica. Some years later, studies of the continental shelf and
slope benthic communities were also developed, while deep-sea benthic community studies were scarce and restricted to the expeditions of the Argentinean Basin
by the RV *Atlantis II* (Bastida, 1973; Bastida *et al.*, in press b).

Although important collections of benthic fauna have been made since the
Challenger expedition in 1876, there is still an important lack of knowledge of

many invertebrate groups. In general there is more interest in Antarctic benthic communities than in those of South America. In the last two decades, Antarctic benthic ecologists have been increasingly concerned with ecology species studies, as well as with trophic relations, recruitment, and population dynamics (Arntz, Brey, and Gallardo, 1994; Clarke, 1996a, 1996b; Favero, Silva, and Ferreyra, 1997; Silva and Favero, 1998).

Most recent studies in the SWAO have focused on the Argentinean continental shelf and its biogeographical relations with Uruguay and Southern Brazil (Bastida, Roux, and Martinez, 1992). On the basis of the benthic assemblages, the southern portion of the SWAO can be divided into three main ecological areas (Figure 12.3). The largest one (area C) includes the outer shelf, influenced by the cold and highly productive waters of the Malvinas-Falklands Current. Between 15 and 20 per cent of its fauna are endemic in this area and correspond perfectly with the Malvinas-Falkland District of the Magellanic biogeographic province (Figure 12.1); this area reaches the highest diversity values based on the Shannon-Wiener and Margalef indices (Bastida, Roux, and Martinez, 1992; Roux et al., 1988).

The second area (B) extends along the Patagonian coast, from Tierra del Fuego to the Península Valdés, and then separates from the coast to the north (Figure 12.3). This area is influenced by the Patagonian Current, and can be considered a transitional zone with lower percentages of species exclusive to it, but taxonomically related with the previous one. It corresponds to the Patagonian District of the Magellanic biogeographic province. The diversity values of this sector of the continental shelf are a little lower than the outer sector, but if the intertidal assemblages are considered it reaches high values in relation to extensive and more diversified habitats, favoured by its high tidal range of about 10 m in many localities.

The third area (A) occupies the inner shelf from Península Valdés to the north (Figure 12.3). This area corresponds to the Bonaerensian District of the Argentinean biogeographic province and is indirectly influenced by the Brazil Current. Between 16 and 20 per cent of the total number of species found here are exclusive to this area. Based on cluster analysis, the species of this area are completely segregated from the species of the two previously mentioned areas. The diversity values of its benthic assemblages are a little higher than those of its neighbouring areas, and also increase when intertidal communities are included.

This basic scheme of the benthic communities of the southern portion of the SWAO extends north, including part of the continental shelf of Uruguay and southern Brazil. Benthic communities are directly influenced by the Brazil Current and taxonomic assemblages of different invertebrate groups differ clearly from those of Magellanic origin, defining the South Brazilian biogeographical province. Although ecological studies in this province are scarce, it seems that benthic biodiversity here increases very little, though this tendency changes towards the boundaries of the Caribbean biogeographical province, where biodiversity is raised to higher values.

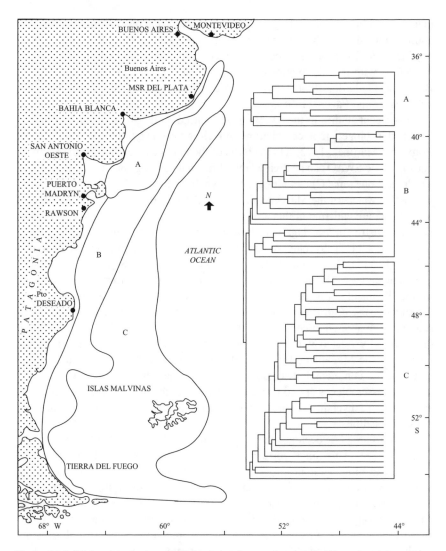

Figure 12.3 Main ecological areas of the Argentine continental shelf based on invertebrate benthic communities

Source: After Bastida, Roux, and Martinez (1992)

Surprisingly, at present there is no convincing evidence of a latitudinal diversity decline in marine shallow-water benthos in the SWAO, as seems to happen in some benthic communities in the northern hemisphere. At any rate, there is still much to learn about many taxonomic groups in the southern hemisphere, and many diversity-related observations are made with little consideration of habitat diversity of the region (Sanders, 1968).

General aspects of SWAO fish assemblages

Studies of fish assemblages and distribution patterns in the SWAO were reported by Menni and López (1984). These investigations were based on the biological collections obtained by Japanese research vessels (*Orient Maru*, *Kaiyo Maru*, and *Shinkai Maru*) in the Argentinean continental shelf at the end of the 1970s.

Based on a total of about 100 fish species sampled along the continental shelf of Argentina and Uruguay, it was possible to separate them into eight different groups with a consistent composition, demonstrating that such groups (or at least most of their species) remain in a definite area during considerable periods of time. New environmental information (temperature, salinity, depth, sediments, etc.) on the fish groups' distribution areas has been verified along different cruises and in different years, giving consistency to the fish associations defined. It is interesting to point out that the results obtained in fish studies are consistent with those obtained in benthic communities, since both fauna delimit in a similar manner the extension of the Magellanic and Argentinean biogeographic provinces.

Fish diversity in the Antarctic Complex shows low values, a situation probably due more to a low habitat diversity than to extreme conditions in the environment. The Subantarctic Complex increases its diversity notably, with a high percentage of commercial species, while the diversity of both provinces of the complex (Magellanic and Argentine) is quite similar but composed of different species.

The Subtropical Complex increases its diversity only moderately in a large part of its extension, except in its northern limit where it is higher due to the presence of coral reef species from the Tropical Complex. The highest diversity of the SWAO corresponds to this complex, where the region has more than 1,000 fish species. As mentioned in the case of benthic communities, there is no clear latitudinal decline for this vertebrate group.

General aspects of the SWAO birds

The coastline of the SWAO is inhabited by more than 300 species of birds, of which about 40 per cent depend on marine habitats to varying degrees, for both breeding and foraging (Table 12.1). These birds can be divided into two main groups: seabirds (i.e. penguins, albatrosses, petrels, gulls, terns) and shorebirds (e.g. plovers and sandpipers). From the former group, penguins can be separately analysed due to their distribution, biomass levels, and foraging strategies.

Seabirds

Due to their coastal and oceanographic characteristics, the SWAO waters are inhabited by a great diversity of seabirds. Despite the oceanic islands (mainly the Fernando de Noronha Archipelago, Brazil) and important harbour colonies,

Table 12.1 Marine bird families (seabirds and shorebirds) that breed in the SWAO

Locality	SUL	FRE	CHA	HAE	LAR	STE	RYN	PHA	SPH	PRO	STR	DIO	HYD	PEL	CHI
Is. Trinidad / M. Vas	✓	✓													
Cabo Frío-R. Janeiro	✓	✓								✓					
Is. Alcatraces	✓	✓	✓		✓	✓		✓							
Is. Castillo – Currais	✓	✓				✓									
Is. Moleques do Sul		✓			✓	✓									
Lagoa do Peixe			✓	✓	✓	✓	✓								
Is. Coronilla			✓	✓	✓	✓									
Lag. Rocha-Is.Lobos					✓										
Is. Flores-Py. Penino				✓	✓	✓		✓							
Punta Rasa			✓	✓	✓	✓									
Mar Chiquita			✓	✓	✓	✓	✓								
A° Zabala			✓	✓	✓	✓									
Bahía Blanca				✓	✓	✓									
Bahía Anegada				✓	✓	✓									
Islote Lobos				✓	✓	✓									
Península Valdés					✓	✓		✓	✓	✓					
Tombo-Bustamante			✓		✓	✓		✓	✓	✓	✓				
Caleta Olivia								✓							
Cabo Blanco															

184

Ría Deseado
Ría Santa Cruz
Cabo Vírgenes
Ba. San Sebastián
Cabo Peñas
Is. de los Estados
Canal de Beagle
Malvinas-Falkland Is.
S.Georgia Is.
S.Sandwich Is.
S.Orkney Is.
Elephant Is.
S.Shetland Is.
Antarctic Peninsula

SUL = *Sulidae*, gannets/boobies; FRE = *Fregatidae*, frigatebirds; CHA = *Charadriidae*, plovers/sandpipers; HAE = *Haematopodidae*, oystercatchers; LAR = *Laridae*, gulls; STE = *Sternidae*, terns; RYN = *Rynchopidae*, skimmers; PHA = *Phalacrocoracidae*, cormorants/shags; SPH = *Spheniscidae*, penguins; PRO = *Procellariidae*, petrels; STR = *Stercorariidae*, skuas; DIO = *Diomedeidae*, albatrosses; HYD = *Hydrobatidae*, storm petrels; PEL = *Pelecanoididae*, diving petrels; CHI = *Chionidae*, sheathbills.

the general distribution of breeding seabirds show an increase in the number of species corresponding to latitude, with about twice the number of species reported breeding in the Malvinas-Falkland and South Georgia Islands than southern Brazil (Figure 12.4). Some groups are restricted to northern localities in Brazil, like boobies and frigatebirds (*Sulidae* and *Fregatidae*), while at higher latitudes there are penguins, albatrosses, and petrels (*Sphenisciformes* and *Procellariiformes*). Gulls, terns, shags, and other minor groups show extended distribution along the coasts.

Breeding periods, considered from the arrival of adults to chick-fledgling, are roughly from late austral winter to mid-autumn, depending on the species and latitude considered (e.g. kelp gulls breed during June to August in Brazil; July to November in Uruguay; October to January in Argentina; and December to February in Antarctica). During the reproductive seasons, foraging areas for seabirds range from 10 to 30 km in smaller species (e.g. terns) to extraordinarily long distances (more than 1,000–2,000 km) for albatrosses and petrels (Prince *et al.*, 1992; Weimerskirch *et al.*, 1993). That means, for example, that an

Figure 12.4 Number of breeding seabird species in the SWAO, and number of non-breeding seabirds in refuelling areas of the SWAO

albatross breeding on South Georgia Islands could be foraging near the Argentinean coast or in the South Shetland Islands of Antarctica.

Penguins

From a total of 17 penguin species around the world, seven regularly breed in South America, with five in the SWAO: king (*Aptenodytes patagonicus*), gentoo (*Pygoscelis papua*), rockhopper (*Eudyptes chrysocome*), macaroni (*E. chrysolophus*), and Magellanic penguins (*Spheniscus magellanicus*) (Del Hoyo, Elliot, and Sargatal, 1992).

Almost all penguin species breed in extensive colonies. Along the Argentine coasts, Magellanic penguins have the largest colonies at Punta Tombo (225,000 pairs) and at Cabo Vírgenes (Del Hoyo, Elliot, and Sargatal, 1992). King penguins in the Malvinas-Falklands have a breeding population of only 400 pairs (Bingham, 1996). King penguins have not bred on the South American mainland since the colony on Isla de los Estados was exterminated by sealers during the last century (Bingham, 1998). The Malvinas-Falkland Islands hold about 20 per cent of the total gentoo penguin population with around 65,000 breeding pairs (Bingham, 1996). Southern rockhopper penguins (*Eudyptes chrysocome chrysocome*) have a distribution restricted to the Malvinas-Falkland Islands (300,000 pairs) and the mainland coast (200,000 pairs) (Bingham, 1998; Frere *et al.*, 1993).

The present Magellanic penguin population for South America and the Malvinas-Falklands is estimated to be around 1.5 million pairs, from which about 650,000 pairs breed along the mainland Argentinean coast (Gandini *et al.*, 1999) and 100,000 pairs in the Malvinas-Falklands (Bingham, 1998). Populations on the Malvinas-Falkland Islands have declined to half their levels in the 1980s. This tendency, however, has not been observed at the Chilean breeding sites of the Strait of Magellan, despite close proximity and similar breeding sites in both localities. The penguin population in the Strait of Magellan has increased during recent seasons, showing chick survival rates to be much higher than those observed in the Malvinas-Falkland colonies. The lower levels of adult mortality or higher recruitment reported for the populations of the Strait of Magellan could be related to the fact that there is no commercial fishing activity around this area. This trend is absolutely different from that observed in other areas of the SWAO (Bingham, 1998; Gandini *et al.*, 1999).

Shorebirds

The coastline of the SWAO must be considered not only as an adequate breeding area, but also as a wintering and/or refuelling area for some long-distance migrant nearctic shorebirds, which breed in the high arctic tundra (i.e. Alaska and Canada) and spend a non-breeding season in South America during the austral summer. Roughly 26 nearctic shorebird species use the shores of the SWAO between

September and April; three other Patagonian breeding species move northwards during the autumn and winter.

The combination of cool ocean currents flowing north along the coast and the geomorphological characteristics of the SWAO shoreline seem to present favourable and productive conditions for shorebirds. The extended beaches of Uruguay and southern Brazil hold important numbers of shorebirds due to the presence of shallow coastal lagoons that are periodically connected to the sea (e.g. Lagoa do Peixe in Brazil and Laguna de Rocha in Uruguay). In Argentina, the most important areas for shorebirds are further south in the Península Valdés (42°S), Golfo San Jorge (46°S), Bahía San Sebastián (53°S), and Bahía Lomas (the eastern mouth of the Strait of Magellan, 52°30'S) in Tierra del Fuego (González, Piersma, and Verkuil, 1996) (Figure 12.4). These coastlines are washed by cool currents rich in nutrients and have high tidal ranges. In the two above-mentioned bays, the conditions of sediment deposition and the intertidal substrata favour the productivity conditions that have led to the development of intertidal areas with invertebrate fauna capable of supporting a high predation pressure of large numbers of shorebirds (Morrison and Ross, 1989).

General aspects of SWAO marine mammals

The marine mammal species of the SWAO comprise 41 cetaceans, four pinnipeds, and one manatee (Table 12.2). Some of them (West Indian manatees, tucuxi, spotted dolphins) inhabit subtropical waters, whereas others (southern elephant seals, hourglass dolphins, beaked whales) have a temperate origin.

Regarding the baleen whales, all balaenopterids are recorded in the SWAO, although most of them correspond to isolated and sporadic stranding specimens. This trend might show that their populations are small in this area; the only

Table 12.2 Marine mammal species in the SWAO

Order	Suborder	Family	Number of species
Cetaceans	Mysticeti	*Neobalaenidae*	1
		Balaenidae	1
		Balaenopteriidae	6
	Odontoceti	*Pontoporiidae*	1
		Iniidae	1
		Phocoenidae	2
		Ziphiidae	7
		Delphinidae	22
Carnivores	Pinnipedia	*Phocidae*	1
		Otariidae	3
Sirenians		*Trichechidae*	1

information about balaenopterid populations derived from commercial catches of fin, sei, minke, and probably Bryde's whales carried out near Cabo Frío (Brazil) until the 1980s. Currently the most important area for the mysticete family is the Abrolhos Bank (17°S/38°W, in Brazil), where a group of around 700 humpback whales breed annually (Siciliano *et al.*, 1990).

In the SWAO, pygmy right whale (*Caperea marginata*) and the southern right whale (*Eubalaena australis*) were recorded. The latter is by far the most important whale in the region, with two breeding concentrations: Península Valdés (42°S, Argentina) and the area of Santa Catarina (28°S, Brazil). During late winter and spring (June to November) approximately 600 whales breed in Península Valdés (Bastida and Lichtschein, 1984b; unpublished data), whereas in southern Brazil smaller groups have been recorded since the 1980s. Both populations are actually increasing at annual rates close to 7 per cent (Bastida and Lichtschein, 1984a; unpublished data). The Península Valdés population has been subject to whale-watching activities during the last 20 years (Lichtschein and Bastida, 1983). The right whale population of the SWAO is one of the most important in the world, comparable to that of South Africa and larger than those of Australia, New Zealand, and subantarctic archipelagos. The southern right whale is specifically protected by law in Argentina, Uruguay, and Brazil, and in the whole region there is a special effort and interest in its conservation. It was made the Natural Monument of Argentina in 1984.

The odontocetes are represented by the families *Physeteriidae* (sperm whales); *Ziphiidae* (beaked whales); *Phocoenidae* (porpoises); *Pontoporiidae* (La Plata dolphin); and *Delphinidae* (marine dolphins). Of these, the latter is the most diverse, and is commonly recorded in the area. Many of the odontocete species have a wide distribution (*Physeter macrocephalus, Orcinus orca, Ziphius cavirostris*), whereas others are strictly restricted to South America (*Cephalorhynchus eutropia, Phocoena spinipinnis, Lagenorhynchus australis*) or even are exclusive to the SWAO (*Pontoporia blainvillei* and *Sotalia fluviatilis*).

Some species clearly show a great latitudinal distribution, being found from the subtropical waters of Brazil to the subantarctic waters of Tierra del Fuego (*Grampus griseus, Physeter macrocephalus, Orcinus orca, Pseudorca crassidens*). Others, in contrast, show clear restrictions to zoogeographical boundaries. Among the coastal species, subtropical and subantarctic ones are equally represented (Figure 12.5) and zoogeographically define three main areas: an area restricted to tropical and subtropical species (<25°/30°S), a transition area (30°/40°S), and a clearly subantarctic area (>40°S) (Figure 12.6). This trend is very similar to that of the continental shelf species, although tropical and subtropical species are better represented, probably due to the southbound extension of the Brazil Current during spring and summer. The dolphins of the genus *Stenella* show the same trend. In the outer continental shelf area this trend is the opposite, because the subantarctic species, mainly beaked whales, are the most commonly represented (Figures 12.5 and 12.6). The presence of subantarctic

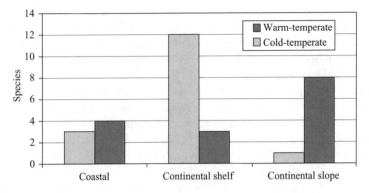

Figure 12.5 Distribution of cold-temperate and warm-temperate cetacean species in the SWAO

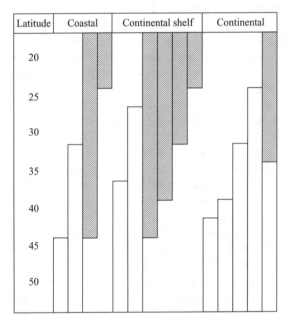

Figure 12.6 Latitudinal distribution of cold-temperate and warm-temperate cetacean species

species in low latitudes may be due to the northern influence of the Malvinas-Falkland Current during winter.

The pinnipeds of the SWAO are much less diverse and are represented by two species (*Otaria flavescens* and *Arctocephalus australis*) of eared seals in the family *Otariidae* and one species (*Mirounga leonina*) of true seal in the family

Phocidae. Both otariids are restricted to South America, and their breeding colonies in the Atlantic are distributed in two main areas, the Patagonian-Fuegian coast (>42°S) (Carrara, 1952) and the Uruguayan coast between 33 and 34°S, where approximately 300,000 animals of both species breed (Vaz Ferreira, 1982a, 1982b). Antarctic fur seal (*Arctocephalus gazella*) breeds on South Georgia Island and the Scotia Archipelago (Bonner, 1979; McCann and Doidge, 1987).

Stock estimates are partial and discontinuous, but some gross population assessments can be summarized (Table 12.3). The population trends vary greatly by region. Sea-lions are steeply decreasing in Uruguay and the Malvinas-Falkland Islands (Hunter *et al.*, 1995; Vaz Ferreira and Ponce de León, 1987), whereas the northern Patagonian population has remained stable during the last 20 years (Crespo and Pedraza, 1991). Fur seals, on the other hand, have been increasing continuously over the last few decades (Vaz Ferreira and Ponce de León, 1987).

The southern elephant seals in the SWAO are restricted to South Georgia and Malvinas-Falkland Islands and to the Península Valdés (the only continental elephant seal breeding ground in the world) (Carrara, 1952). These colonies seem to belong to the same population stock as the colonies of Gough, South Orkney, and South Shetland Islands (Laws, 1994); this stock, which comprises about 50 per cent of the world population, has about 400,000 animals, a number that has remained stable or has slightly increased over the last 50 years (Boyd, Walker, and Poncet, 1996); the only exception is the expanding colony of the Península Valdés (Campagna and Lewis, 1992).

During recent years there has been an increasing number of seal vagrants recorded in the SWAO, belonging to species that do not breed in South America. That is the case for Antarctic fur seals (*Arctocephalus gazella*), leopard seals (*Hydrurga leptonyx*), crabeater seals (*Lobodon carcinophagus*), and Weddell seals (*Leptonychotes weddellii*). Much more frequently recorded are subantarctic fur seals (*Arctocephalus tropicalis*), which surely come from the breeding colonies of the Gough Islands (Rodríguez, 1996). The increasing number of these vagrants indicates not only the influence of large-scale oceanic currents, but may also be a consequence of their increasing population size.

Table 12.3 Pinnipeds stock estimates for the SWAO

Region	Sea-lions	Breeding colonies	Fur seals	Breeding colonies
Uruguay	15,000	6	250,000	6
Continental Argentina	70,000	70	10,000	12
Malvinas-Falkland Islands	6,000	65	-	16
Total	106,000	141	260,000	34

Main environmental and conservation problems of the region

Along the SWAO there are certain areas where the planktonic assemblages play a key role in relation to the reproductive success of commercial fishes and invertebrates (e.g. Golfo San Jorge (46°S) in Patagonia and Bahía Samborombón (36°S) in the Buenos Aires Province). These areas are of special interest in cases of severe impact, as in the event of an oil spill, because this event could directly affect the success of fish populations.

Conservation problems with benthic invertebrates are related mainly to the overexploitation of several species, as in the case of the mussel (*Mytilus edulis platensis*), the yellow clam (*Mesodesma mactroides*), the scallop (*Chlamys tehuelchus*), and the king crab (*Lithodes santolla*). Many benthic communities are affected as a result of the by-catch during bottom commercial fishing, which impacts on sediment stability (Bastida *et al.*, in press b). The shrimp (*Pleoticus muelleri*) fishery, of high commercial value in northern Patagonia (Boschi, 1986), is responsible for high impact by-catches, affecting mainly juveniles hake and Magellanic penguins (Gandini *et al.*, 1999). Another interesting case of invertebrate conservation is the massive death of yellow clams along the SWAO during the spring of 1995. For unknown causes the clam populations of southern Brazil, Uruguay, and Argentina drastically declined in a couple of months, and no adult specimens were found during the following years. Currently only a small adult stock survives (near Bahía San Blás, 40–30°S, Argentina) and the potential for recovery is low (Bastida *et al.*, in press a).

The intertidal and subtidal benthic habitats around many cities and harbours of the region are often affected by various human activities (sewage dumping, industrial effluents, coastal constructions, introduction of alien species, etc.). There is also a potential risk for shelf benthic communities with the increment in offshore oilrigs.

Usually, conservation problems with fishes and other nektonic species such as squids are related to overfishing. The SWAO fishing grounds, most importantly those of Uruguay and Argentina, have faced dramatic changes in the last decades. Although many SWAO fishery resources were unexploited or underexploited during the 1960s, during the 1990s there has been a severe impact on SWAO hawkfish (*Cheilodactylus bergi*) stocks, due to the activities of Soviet fishing fleets.

The SWAO is inhabited by around a thousand fish species, only 50 species of which are of true commercial value and have been assessed. To these commercial resources some highly valued invertebrate species may be added, as in the case of squids. Recent stock assessments figure a total biomass from these 50 commercial species reaching about 14 million tonnes. About 80 per cent of this biomass is in four species: anchovy (*Engraulis anchoita*), hake (*Merluccius hubbsi*), longtail hake or hoki (*Macruronus magellanicus*), and shortfin squid (*Illex argentinus*) (Figure 12.7).

The fisheries of the three SWAO countries catch around 2 million tons annually. Figure 12.8 shows an increase in the total catch from the end of the 1970s up to 1995 and despite increased fishing effort, followed by a clear decline. There are now clear signs that many species are being overexploited. The most

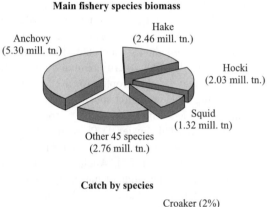

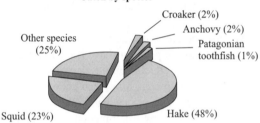

Figure 12.7 Estimated biomass and fishery catch of the main species in the SWAO

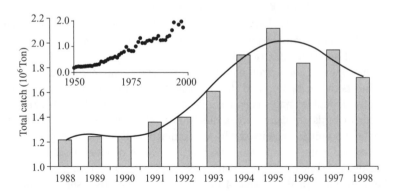

Figure 12.8 Catch fluctuation of fisheries during 1988–1998 in the SWAO, and the historical trend over 50 years

important is the hake, followed by the Whitemouth croaker (*Micropogonias furnieri*). Both species maintain smaller biomass levels but are, nevertheless, extremely important for coastal ecosystems.

Taking into account several trophic web studies carried out in the SWAO, it is clear that anchovy, hake, and shortfin squid are the key species of the region (Angelescu, 1982). Many fish and higher vertebrate species (marine birds and mammals) rely on this trophic tripod. Hake is heavily overexploited, and regulatory measures taken until now do not seem to have established a more rational exploitation scheme or a rehabilitation of this resource, which is of high international demand. The shortfin squid shows wide natural biomass variations due to its short life span, and if periods of low recruitment and intense fishing pressure combine, the consequences may be catastrophic for the resource and the species relying on it. Anchovy is the most abundant species, with a total biomass higher than 5 million tonnes, divided into two fishing stocks, the northern and Patagonian (Brandhorst *et al.*, 1974; Fuster de Plaza and Boschi, 1958; Hansen, Cousseau, and Gru, 1984). Most commercially important fish depend on this small species for food (Angelescu, 1982). Anchovy is underexploited at the moment, but due to the explosive demands of the international fishery market it must be properly managed, with exploitation levels kept lower than the proposed 140,000 tonnes/yr.

The main conservation problems for seabirds and shorebirds are of a different origin. Due to their inability to fly, penguins have been a very easy food resource for humans. This had harmful effects during the nineteenth century. In the Malvinas-Falkland Islands penguin eggs have been collected since 1700, and this exploitation ceased recently. In other places, egg collecting is still a local threat. This refers to the egging carried out by fisherman on tern colonies in southeastern Brazil, and to the collecting of gulls' eggs in Uruguay and Argentina (Del Hoyo, Elliot, and Sargatal, 1996; Escalante, 1991; Zuquim Antas, 1991; unpublished data).

Another reason for penguin exploitation has been oil extraction. Between 1864 and 1880, about 2,500,000 penguins (mostly rockhoppers) were killed in the Malvinas-Falklands for their oil (Martínez, 1992). During the 1940s, about 40,000 Magellanic penguins were caught annually on the Argentine coast for the skin and oil industries (Carrara, 1952).

The more severe conservation problems for seabirds and marine birds today are related to both the fishing and the petroleum industries. Incidental mortalities, especially in boobies and shags, have been significant in southern Brazil, where hundreds of birds are caught in sardine (*Sardinella*) nets. In subantarctic areas, incidental mortalities associated with the commercial fishery of Patagonian toothfish have had a great impact on the mortality rates of albatrosses (Alexander, Robertson, and Gales, 1997; Croxall and Wace, 1992). Overfishing of sardine and anchovy in southern Brazil and of the above-mentioned fish, shrimp, and squid stocks in Uruguay and Argentina has reduced available food for seabirds (Escalante, 1991). Direct competition with commercial fisheries also occurs in

Antarctica and subantarctic areas, mainly in relation to *Nototheniid* and *Chaenictid* species (Croxall *et al.*, 1984).

Oil spills have potential importance in colonies located in the states of Sao Paulo and Rio de Janeiro in Brazil, where major oil terminals are located. In Uruguay and Argentina several oil spills have occurred with local but severe impact on marine communities. Taking into account the breeding phenology, the period with greatest risk is between September and May, when not only most of the species are breeding, but also important nearctic and austral migratory populations can be found on the shoreline. The Magellanic penguin is the species mostly affected by oil spills in the SWAO, especially during their migration period. Studies performed on the Patagonian coast and in rehabilitation centres in north Argentina, Uruguay, and Brazil allow us to make a rough annual mortality estimation of 70,000 to 120,000 individuals (Bastida *et al.*, 1998; Gandini *et al.*, 1994). The intensive maritime traffic in the Rio de la Plata estuary, and the presence of two important commercial harbours at Buenos Aires and Montevideo, led to the habitat loss of the surroundings.

In South America, important areas for shorebirds are threatened by agricultural and pesticide pollution, industrial and port developments, forestry, shrimp farming in southern Brazil, tourism in northern Argentina, and oil development in Tierra del Fuego. Although some of these developments could potentially lead to catastrophic effects, the result of "creeping erosion" on a smaller scale must also be considered (Morrison and Ross, 1989). Due to their very narrow distribution and small population size, declining status, or unfavourable population viability, some species of the SWAO coasts, such as red-legged cormorant (*Phalacrocorax gaimardi*), Magellanic plover (*Pluvianellus socialis*), dolphin gull (*Leucophaeus scoresbi*), and Olrog's gull (*Larus atlanticus*) have been considered vulnerable.

Marine mammals have been exploited for centuries around the world, and the SWAO has been no exception. Apart from the well-known exploitation of whales, pinnipeds were commercially harvested in the SWAO for decades. Fur seal exploitation in Uruguay began in the sixteenth century and ended in 1991 with an estimated total catch of 750,000 animals (Vaz Ferreira and Ponce de León, 1987). Sea-lion exploitation came later, with the main harvests conducted in Uruguay, Patagonia, and the Malvinas-Falkland Islands during this century (Carrara, 1952; Vaz Ferreira, 1982a). The total number of animals harvested was also about 750,000. The Antarctic fur seal is probably one of the most irrationally harvested pinnipeds in the world. This species has come very close to extinction at least twice, with an estimated 1,200,000 animals killed from the eighteenth to the nineteenth centuries (McCann and Doidge, 1987; Weddell, 1825). Over the last 50 years, the recovery of this species has been spectacular, and the population currently exceeds 2 million animals (Boyd, 1993). From the end of the eighteenth century to the beginning of the nineteenth, elephant seal catches were conducted in the Malvinas-Falkland and South Georgia Islands, which put the stocks in a critical condition. After a brief period of recovery, controlled harvests were conducted during the twentieth century untill the 1960s (Laws, 1994). During that century, elephant sealing was also conducted at Península Valdés in northern Patagonia (Carrara, 1952).

The marine mammals of the SWAO have faced several conservation problems during recent decades. Among them, by-catch of the commercial fisheries has been the most important. In southern Brazil, Uruguay, and northern Argentina, coastal gillnets are responsible for high incidental mortality rates for La Plata dolphins (*Pontoporia blainvillei*) (Brownell, 1989; Pinedo, Praderi, and Brownell, 1989; Praderi, 1997; Praderi, Pinedo, and Crespo, 1989). Although there are no assessments for the whole area, about 1,500 to 2,000 dolphins are captured annually, presenting a serious problem regarding the size, trend, and identity of the stocks. In this type of fishery, small amounts of other marine mammal species (sea-lions, Burmeister's porpoises, dusky and bottlenose dolphins) are also taken. Trawl fishing in northern Patagonia also produces an incidental catch of dusky dolphins, Commersonn's dolphins, and sea-lions, and their numbers of annual catch were estimated at 70–200, 25–170, and 170–480 individuals, respectively (Crespo *et al.*, 1997).

There is also some suspicion of direct resource competition between fisheries in the area, as evidenced by overfishing in the Uruguay and Patagonia grounds. The Malvinas-Falkland stock of sea-lions has drastically decreased in recent decades, diminishing by more than 90 per cent (Hunter *et al.*, 1995). Although there are no clear causes for this decline, competition between fisheries in the area has been suggested. In recent years there has been no record of marine mammal exploitation, though during the 1980s and early 1990s Commersonn's dolphins were used in Tierra del Fuego and southern Chile as bait in king crab (*Lithodes santolla*) traps (Goodall *et al.*, 1988).

Marine mammals are legally protected in Brazil, Uruguay, and Argentina and, as previously stated, only Southern right whales in the Península Valdés are exploited by regulated whale-watching activities. Small-scale tourist activities are also taking place in Brazil, centring on the spinner dolphin populations of the Fernando de Noronha Archipelago. These activities may produce negative effects in the near future if legal regulations are not strictly reinforced.

The SWAO has received relatively little attention from an environmental point of view. Only a few pollution studies have been carried out from the middle of 1970s. Most of the studies have been undertaken in heavily populated areas, or areas with petrochemical or potentially hazardous industries. Pollution studies, such as heavy metals and organochlorine compounds, were made more actively compared with other contaminant studies. These studies were mainly on coastal ecosystems (e.g. Seelinger, de Lacerda, and Patchineelam, 1988) and their various faunistic groups and trophic webs. Although no regional assessment programme exist, gross results indicate that high values of heavy metals can be found in the seas, sediments, and biota throughout South America, and some of these values exceed widely held standards for human consumption (Nauen, 1983).

Table 12.4 Mean mercury, cadmium, and lead concentration (ppm) in several marine groups of the SWAO

Group	Number of species	Mercury	Cadmium	Lead	Source
Algae	8 (AN)	nd	0.05–2.01 (W)	—	1
Porifera	1 (AN)	nd	3.48–3.84 (W)	—	1
Brachiopods	1 (AN)	nd	3.70–4.95 (W)	—	1
Briozoa	2 (AN)	nd	0.01–1.01 (W)	—	1
Polychaeta	1 (AN)	nd	0.12–0.21 (W)	—	1
Molluscs (mussels)	2 (SA)	0.05–0.60 (M)	1.6–6.7 (M)	1.6–18.0 (M)	2, 3, 4, and 5
Molluscs (limpets)	1 (AN)	nd	0.21–3.75 (M)	—	1
Molluscs (snails)	1 (AN)	nd	1.36–3.75 (M)	—	1
Crustaceans (shrimps)	2 (SA)	0.11–0.16 (M)	0.03–0.05 (M)	—	6 and 7
Crustaceans (benthic isopods)	3 (AN)	nd	0.98–1.91 (W)	—	8
Crustaceans (planktonic amphipods)	1 (AN)	nd	8.00–118.0 (W)	—	9 and 10
Elasmobranchs	4 (SA)	0.10–2.45 (M)	ND–0.15 (M)	—	6 and 7
Teleost	12 (SA)	0.05–0.73 (M)	0.60–3.70 (M)	—	6 and 7
Teleost	1 (AN)	0.01–0.10 (M)	<0.05 (M)	—	1
Pinnipeds	5 (SA)	0.60–121.1 (L)	0.60–48.60 (K)	1.44–1.79 (B)	11 and 12
Pinnipeds	1 (AN)	4.10–7.60 (L)	3.70–5.90 (K)	—	1
Cetaceans	12 (SA)	0.01–63.7 (L)	nd–412.60 (K)	—	11

L = Liver; M = Muscle; W = Whole body; K = Kidney; B = bone; ND = non-detectable; SA = South America; and AN = Antarctica.

Sources: Moreno et al., 1997 (1); Bernasconi et al., 1991 (2); Scarlato and Marcovecchio, 1996 (3); Batalha and Wallner Kersanach, 1996 (4); Gil et al., 1997 (5); Marcovecchio, Moreno, and Perez, 1988 (6); Marcovecchio, Moreno, and Perez, 1989 (7); Petri and Zauke, 1993 (8); Schulz-Baldes, 1992 (9); Rainbow, 1989 (10); Gerpe, 1996 (11); Peña et al., 1988 (12)

In the Atlantic sector of Antarctica, heavy metals are present in biota, albeit in lower values (Moreno *et al.*, 1997). Selected values for mercury, cadmium, and lead in South America and Antarctica are included in Table 12.4.

Information about the presence of organochlorine compounds (such as organochlorine pesticides and polychlorinated biphenyls) in the SWAO is more limited. These studies also showed the presence of these contaminants in water, sediments, and biota, mainly in areas of extensive agricultural and industrial development, such as in the Río de la Plata estuary in Argentina and Uruguay (Colombo *et al.*, 1990; Gil *et al.*, 1997; Janiot, Sericano, and Roses, 1994; Lanfranchi *et al.*, 1998). Finally, the potential risk of nuclear contamination in the Malvinas-Falkland Islands, as a consequence of the 1982 war, should be noted. During the conflict, the warships *Sheffield*, *Ardent*, *Antelope*, and *Coventry* were destroyed and sunk by Argentinean forces and, although officials deny that these warships were carrying nuclear weapons, British MPs revealed to the press that classified documents indicated nuclear weapons were transported to the conflict area. This should be confirmed one way or the other, and the possibility of nuclear contamination in the SWAO completely ruled out.

ACKNOWLEDGEMENTS

The authors sincerely thank Professor Dr Nobuyuki Miyazaki (Otsuchi Marine Research Center, Ocean Research Institute, University of Tokyo, Japan) for his kind invitation to participate in the International Conference "Man and the Ocean" and for his friendship. The Agencia Nacional de Promoción Científica y Tecnológica of Argentina gave financial support to the present study (PICT97 # 07-00000-01651) and Cart. Marcelo Farenga (Centro de Geología de Costas y del Cuaternario, Universidad Nacional de Mar del Plata, Argentina) kindly assisted with the illustrations.

REFERENCES

Alexander, K., G. Robertson, and R. Gales. 1997. The incidental mortality of Albatrosses in longline fisheries. *Report of the First International on the Biology and Conservation of Albatrosses*, 42pp. Hobart (Australia).

Angelescu, V. 1982. Ecología trófica de la anchoíta del Mar Argentino (Engraulidae, *Engraulis anchoita*). Parte II. Alimentación, comportamiento y relaciones troficas en el ecosistema. *Contribuciones del Instituto Nacional de Investigación y Desarrollo Pesquero*, 409: 83pp.

Arntz, W.E., T. Brey, and V.A. Gallardo. 1994. Antarctic zoobenthos. *Oceanography and Marine Biology Annual Review*, 32: 241–304.

Balech, E. 1964. Caracteres biogeográficos de Argentina y Uruguay. *Boletín del Instituto de Biología Marina*, 7: 107–112.

Balech, E. 1976. La distribución de algunos micropláncteres en el Atlántico Sur Oeste. *Boletín del Servicio de Hidrografía Naval*, 12(2, 3): 15–24.

Bastida, R. 1973. Sobre el hallazgo de Crustacea Cephalocaridea frente a las costas argentino-uruguayas. *Physis*, A 32(84): 220.

Bastida, R. and V. Lichtschein. 1984a. Estado actual del conocimiento de la ballena franca austral (*Eubalaena australis*) en el Hemisferio Sur. *Resúmenes de la Primera Reunión de Expertos en Mamíferos Acuaticos de América del Sur*: p. 4.

Bastida, R. and V. Lichtschein. 1984b. Informe preliminar sobre los estudios de Ballena Franca Austral (*Eubalaena australis*) en la zona de Península Valdés. *Actas de la Tercera Reunión IberoAméricana de Conservación y Zoología de Vertebrados. Revista del Museo Argentino de Ciencias Naturales Bernardino Rivadavia*, Zoología, Tomo XIII, 21: 197–210.

Bastida, R. and C.M. Urien. 1981. Investigaciónes sobre comunidades bentónicas. Características generales del sustrato (Campanas IV, V, X y XI del B/I 'Shinkai Maru'), pp. 318–339. In: Angelescu, V. (ed.), *Campañas de investigación pesquera realizadas en el Mar Argentino por los B/I Shinkai Maru y Walter Herwig y el B/P Malburg, anos 1978 y 1979. Resultados de la parte argentina. Contribuciones del Instituto Nacional de Investigación y Desarrollo Pesquero*, 383.

Bastida, R., A. Roux, and D. Martinez. 1992. Benthic communities of the Argentine continental shelf. *Oceanologica Acta*, 15(6): 687–698.

Bastida, R., S.G. Moron, A. D'Amaral, D. Rodríguez, and L. Rivero. 1998. Fundación Mundo Marino: Memoria Quinquenal 1991–1995. *Fundación Mundo Marino*, 332pp. San Clemente del Tuyu, Argentina.

Bastida, R., E. Ieno, J.P. Martin, and E. Mabragana. (In press a). The yellow clam (*Mesodesma mactroides*): a case study of a coastal resource in risk from the Southwestern Atlantic Ocean. *Journal of Medical and Applied Malacology*.

Bastida, R., M. Zamponi, C. Bremec, A. Roux, R. Elias, and G. Genzano. (In press b). Las comunidades bentónicas del Mar Argentino., In: *El Mar Argentino y sus Recursos Pesqueros*, E. Boschi (ed.), pp. 1–53. Instituto Nacional de Investigación y Desarrollo Pesquero, Mar del Plata, Argentina.

Batalha, F. and M. Wallner Kersanach. 1996. Monitoramento biológico; analise da concentracao de metais pesados em moluscos bivalves na Baia de Sepetiba, Rio de Janeiro. *Resúmenes del Sexto Congreso Latinoamericano de Ciencias del Mar*, p. 29.

Bernasconi, G.A., R. Odino, B. Souto, and S. Mendez. 1991. Estudio del contenido elemental en moluscos de la costa uruguaya por fluorescencia de rayos X. *Comunicaciones de las Jornadas de Investigación Científica en Materia de Contaminación de las Aguas (CTMFM-CARU-CARP)*, pp. 147–153.

Bingham, M. 1996. Penguin population census 1995–6. *The Warrah*, 10: 6–7.

Bingham, M. 1998. The penguins of South América and the Falkland Islands. *Penguin Conservation*, 11: 8–15.

Boltovskoy, D., (ed.). 1981. *Atlas del Zooplancton del Atlántico Sudoccidental*, 936pp. Publicación Especial de Instituto Nacional de Investigación y Desarrollo Pesquero, Mar del Plata, Argentina.

Boltovskoy, E. 1964. Provincias zoogeográfias de América del Sur y su sector Antártico según los foraminíferos bentónicos. *Boletín del Instituto de Biología Marina*, 7: 94–99.

Bonner, W.N. 1979. Antarctic (Kerguelen) fur seal. *FAO Mammals in the Seas*, II: 49–51.

Boschi, E.E. 1986. La pesqueria del langostino del litoral patagonico. *Redes*, 20: 21–26.

Boyd, I.L. 1993. Pup production and distribution of breeding Antarctic fur seals *Arctocephalus gazella* at South Georgia. *Antarctic Science*, 5: 17–24.

Boyd, I.L., T.R. Walker, and J. Poncet. 1996. Status of the southern elephant seals in South Georgia. *Antarctic Science*, 5: 237–244.

Brandhorst, W., J.P. Castello, M.B. Cousseau, and D.A. Capezzani. 1974. Evaluación de los recursos de anchoíta (*Engraulis anchoita*) frente a la Argentina y Uruguay. VII: Desove, crecimiento, mortalidad y estructura de la población. *Physis*, A33(86): 37–58.

Brownell Jr, R.L. 1989. Franciscana, *Pontoporia blainvillei* (Gervais & Dorbigny, 1844). In: *Handbook of Marine Mammals*, S.H. Ridgway and R.J. Harrison (eds), Vol. 4, pp. 45–67. Academic Press, London (UK).

Campagna, C. and M. Lewis. 1992. Growth and distribution of a southern elephant seal colony. *Marine Mammal Science*, 8(4): 387–396.

Carcelles, A. and S. Williamson. 1951. Catálogo de los Moluscos Marinos de la Provincia Magallánica. *Revista del Instituto Nacional de Investigación en Ciencias Naturales, Zoología*, 2(5): 225–383.

Carrara, I.S. 1952. *Lobos marinos, pinguinos y guaneras del litoral maritimo e islas adyacentes de la República Argentina*, 191pp. Catedra de Higiene e Industrias, Facultad de Ciencias Veterinarias, Universidad Nacional de La Plata., La Plata, Argentina.

Carreto, J.I., R.M. Negri, H.R. Benavides, and R. Akselman. 1985. Toxic dinoflagellate blooms in the Argentine Sea. In: *Toxic Dinoflagellates*, D.M. Anderson, A.W. White, and D.G. Baden (eds), pp. 147–152. Elsevier, New York (United States).

Clarke, A. 1996a. The distribution of Antarctic marine benthic communities. *Antarctic Research Series*, No. 70, pp. 219–230.

Clarke, A. 1996b. Marine benthic populations in Antarctica: Patterns and proccesses. *Antarctic Research Series*, No. 70, pp. 373–388.

Colombo, J.C., M.F. Khalil, M. Arna, and A.C. Hort. 1990. Distribution of chlorinated pesticides and individual polychlorinated biphenyls in biotic and abiotic compartments of the Rio de la Plata, Argentina. *Environmental Science and Technology*, 24: 498–505.

Crespo, E.A. and S.N. Pedraza. 1991. Estado actual y tendencia de la población de lobos marinos de un pelo (*Otaria flavescens*) en el litoral norpatagonico. *Ecología Austral*, No.1, pp. 87–95.

Crespo, E.A., S.N. Pedraza, S.L. Dans, M. Koen Alonso, L.M. Reyes, N.A. Garcia, M. Coscarella, and A.C.M. Schiavini. 1997. Direct and indirect effects of the high-seas fisheries on the marine mammal populations in northern and central Patagonian coast. *Journal of Northwest Atlantic Fishery Science*, 22: 189–207.

Croxall, J.P. and N. Wace. 1992. Interaction between marine and terrestrial ecosystem. In: *Progress in Conservation of the Subantarctic Islands. Proceedings of a SCAR/IUCN Workshop on Protection, Research and Management of Subantarctic Islands*, P.R. Dingwall, P.G.H. Evans, and R.W. Schreiber (eds), pp. 115–120. SCAR/IUCN, Cambridge (UK).

Croxall, J.P., P.A. Prince, I. Hunter, S.J. McInness, and P.G. Copestake. 1984. The seabirds of the Antarctic Peninsula, islands of the Scotia Sea, and Antarctic continent between 80W and 20W: Their status and conservation. In: *Status and Conservation of World's Seabirds*, J.P. Croxall, P.G.H. Evans, and R.W. Schreiber (eds), pp. 637–666. ICBP Technical Publication 2, Cambridge (UK).

Del Hoyo, J., A. Elliot, and J. Sargatal (eds). 1992. *Handbook of the Birds of the World, Vol. 1: Ostrich to Ducks*, 696pp. Lynx Ediciones, Barcelona (Spain).

Del Hoyo, J., A. Elliot, and J. Sargatal. 1996. *Handbook of the Birds of the World, Vol. 3: Hoatzin to Auks*, 821pp. Lynx Ediciones, Barcelona (Spain).

Dell, R.K. 1972. Antarctic benthos. *Advances in Marine Biology*, 10: 1–216.

El-Sayed, S.Z. 1968. Prospects for primary productivity in Antarctic waters. *Proceedings of a Symposium on Antarctic Oceanography (SCAR, SCOR, IAPO and IUBPS)*, pp. 227–239.

Escalante, E. 1991. Status and conservation of seabirds breeding in Uruguay. In: *Seabird Status and Conservation: A Supplement*, J.P. Croxall (ed.), pp. 159–164. ICBP Technical Publication 11, Cambridge (UK).

Favero, M., P. Silva, and G. Ferreyra. 1997. Trophic relationship between the kelp gull and the Antarctic limpet at King George Island (South Shetland Island, Antarctica) during the breeding season. *Polar Biology*, 17: 431–436.

Ferguson-Wood, E.J. and J.J. Walsh. 1968. Fluorescence studies of the vertical distribution of phytoplacton studies and studies of species/depth in Antarctic waters. In: *Proceedings of a Symposium on Antarctic Oceanography (SCAR, SCOR, IAPO and IUBPS)*, pp. 41–42.

Frere, E.M., P. Gandini, P. Gandini, T. Holik, V. Lichtschein, and M.O. Day. 1993. Variación anual del número de adultos reproductivos en una nueva colonia de Pinguino Penacho Amarillo en Isla Pinguino (Santa Cruz, Argentina). *Hornero*, 13: 293–294.

Fuster de Plaza, M.L. and E.E. Boschi. 1958. Estudio biológico pesquero de la anchoíta (*Engraulis anchoita*) de Mar del Plata. *Departamento de Investigación Pesquera de la Secretaria de Agricultura y Ganaderia, Publicación* No. 7, pp. 1–49.

Gandini, P., P.D. Boersma, E. Frere, M. Gandini, T. Holik, and V. Lichtschein. 1994. Magellanic penguins (*Spheniscus magellanicus*) afected by chronic petroleum pollution along the coast of Chubut, Argentina. *The Auk*, 111: 20–27.

Gandini, P., E. Frere, A. Pettovello, and P. Cedrola. 1999. Interaction between Magellanic penguins and shrimp fisheries in Patagonia, Argentina. *The Condor*, 101: 783–789.

Gerpe, M. 1996. Distribución y dinámica de metales pesados en mamíferos marinos. PhD *tesis,* Universidad Nacional de Mar del Plata, Argentina, 349pp.

Gil, M.N., M.A. Harvey, H. Beldoménico, S. García, M. Commendatore, P. Gandini, E. Frere, P. Yorio, E. Crespo, and J.L. Esteves. 1997. Contaminación por metales y plaguicidas organoclorados en organismos marinos de la zona costera patagónica. *Informes Técnicos del Plan de Manejo Integrado de la Zona Costera Patagónica*, 32: 28pp.

González, P., T. Piersma, and Y. Verkuil. 1996. Food, feeding and refueling of red Knots during northbound migration at San Antonio Oeste, Río Negro, Argentina. *Journal of Field Ornithology*, 67: 575–591.

Goodall, R.N.P., A.R. Galeazzi, S. Leatherwood, K.W. Miller, P.S. Cameron, R.K. Kastelein, and A.P. Sobral. 1988. Studies of Commerson's dolphins, *Cephalorhynchus commersonii*, off Tierra del Fuego, 1976–1984, with reviews of information in the species in the South Atlantic. In: *Biology of the Genus Cephalorhynchus*, R.L. Brownell Jr and G.P. Donovan (eds), pp. 3–70. International Whaling Commission Special Issue 9.

Hansen, J.E., M.B. Cousseau, and D.L. Gru. 1984. Características poblacionales de la anchoíta (*Engraulis anchoita*) del Mar Argentino. Parte I. El largo medio al primer año de vida, crecimiento y mortalidad. *Revista de Investigación y Desarrollo Pesquero*, 4: 21–48.

Hedgpeth, J.W. 1970. Marine biogeography of the Antarctic regions. In: *Antarctic Ecology*, 1: 97–104. Academic Press, New York.

Hunter, C.J., D. Thompson, C.D. Duck, and M. Riddy. 1995. Status of southern sea-lions in the Falkland Islands. In: *Abstracts of the Eleventh Biennal Conference on the Biology of Marine Mammals*, p. 56.

Janiot, L.J., J.L. Sericano, and O.E. Roses. 1994. Chlorinated pesticide occurence in the Uruguay River (Argentina-Uruguay). *Water, Air and Soil Pollution*, 76: 323–331.

Lanfranchi, A.L., J.E.D. Moreno, V.J. Moreno, T. Metcalfe, and M.L. Menone. 1998. Distribution of organochlorine compounds in tissues of the Croaker (*Micropogonias furnieri*) from Samborombón Bay, Argentina. *Environmental Sciences*, 6(1): 55–67.

Laws, R.M. 1994. History and present status of southern elephant seal populations, In: *Elephant Seals: Population Ecology, Behavior and Physiology*, B. Le Boeuf, and R.M. Laws (eds), pp. 49–65. University of California Press, Santa Cruz.

Lichtschein, V. and R. Bastida. 1983. Whale watching in Argentina. *Proceedings of the Global Conference on the Non-Consumptive Utilization of Cetacean Resources*, pp. 1–4. Boston.

Marcovecchio, J., V.J. Moreno, and A. Perez. 1988. Determination of heavy metal concentration in the biota of Bahía Blanca, Argentina. *Science of the Total Environment*, 75: 181–190.

Marcovecchio, J., V.J. Moreno, and A. Perez. 1989. Cadmium, zinc and total mercury distribution in organisms from Samborombón Bay (La Plata River Estuary) in Argentina. In: *Heavy Metals in the Environment*, J.P. Vernet (ed.), Vol. 1: 366–369. CEP Consultant Ltd., Edinburgh (UK).

Martínez, I. 1992. Family *Spheniscidae* (Penguins). In: *Handbook of the Birds of the World, Vol. 1: Ostrich to Ducks*, J. Del Hoyo, A. Elliot, and J. Sargatal (eds.), pp. 140–160. Lynx Ediciones, Barcelona.

McCann, T.S. and D.W. Doidge. 1987. Antarctic fur seal, *Arctocephalus gazella*, In: *Status, Biology and Ecology of Fur Seals*, J.P. Croxall, and R.L. Gentry (eds), pp. 5–8. NOAA Technical Report NMFS 511.

Menni, R. and H. López. 1984. Distributional patterns of argentine marine fishes. *Physis*, A(103): 71–85.

Menni, R., R. Ringuelet, and R. Arámburu. 1984. *Peces marinos de la Argentina y Uruguay*, 359pp. Editorial Hemisferio Sur, Buenos Aires.

Moreno, J.E.D., M.S. Gerpe, V.J. Moreno, and C. Vodopivez. 1997. Heavy metals in Antarctic organisms. *Polar Biology*, 17: 131–140.

Morrison, R.I.G. and R.K. Ross. 1989. *Atlas of Neartic Shorebirds on the Coast of South America*, Vols. 1 & 2, 325pp. Canadian Wildlife Service Special Publication, Ottawa.

Nauen, C. 1983. *Compilation of Legal Limits for Hazardous Substances in Fish and Fishery Products. FAO* Fisheries Circular, No. 764, pp. 102.

Peña, N.I., V.J. Moreno, J.E. Marcovecchio, and A. Perez. 1988. Total mercury, cadmium and lead distribution in tissues of the southern sea-lion (*Otaria flavescens*) in the ecosystem of Mar del Plata, Argentina. In: *Metals in Coastal Environments of Latin América*, U. Seelinger, L.D. De Lacerda, and S.R. Patchineelam (eds), pp. 140–146. Springer-Verlag, Berlin.

Peterson, R.G. and L. Stramma. 1991. Upper-level circulation in the South Atlantic Ocean. *Progress in Oceanography*, 26: 1–73.

Petri, G. and C.P. Zauke. 1993. Trace metals in Crustaceans, in the Antarctic Ocean. *AMBIO*, 22: 529–536.

Pinedo, M.C., R. Praderi, and R.L. Brownell Jr. 1989. Review of the biology and status of the franciscana. In: *Biology and Conservation of River Dolphins*, W.F. Perrin, R.L. Brownell Jr, and K. Zhou (eds), pp. 46–51. Occasional Papers of the IUCN Species Survival Commission, No. 3.

Praderi, R. 1997. Analisis comparativo de estadisticas de captura y mortalidad incidental de Pontoporia blainvillei en Uruguay durante 20 anos. In: *Anais do Segundo Encontro sobre Coordenacao de Pesquisa e Manejo da Franciscana*, M.C. Pinedo, and A.S. Barreto (eds), pp. 42–53. Fundacao Universidade do Rio Grande, Rio Grande do Sul, Brazil.

Praderi, R., M.C. Pinedo, and E.A. Crespo. 1989. Conservation and management of *Pontoporia blainvillei* in Uruguay, Brazil and Argentina. In: *Biology and Conservation of River Dolphins*, W.F. Perrin, R.L. Brownell Jr, and K. Zhou (eds), pp. 52–56. Occasional Papers of the IUCN Species Survival Commission, No. 3.

Prince, P.A., A.G. Wood, T. Barton, and J.P. Croxall. 1992. Satellite tracking of wandering albatrosses (*Diomedea eluxans*) in the South Atlantic. *Antarctic Science*, 4: 31–36.

Rainbow, P.S. 1989. Copper, cadmium and zinc concentrations in oceanic amphipod and euphausiid crustaceans, as a source of heavy metals to pelagic seabirds. *Marine Biology*, 103: 513–518.

Rodríguez, D. 1996. Biología y ecologia de los pinnipedos del sector bonaerense. PhD tesis, Universidad Nacional de Mar del Plata, Argentina, 351pp.

Roux, A., R. Bastida, V. Lichtschein, and A. Barreto. 1988. Investigaciónes sobre las comunidades bentónicas de plataforma a través de una trasecta frente a Mar del Plata. *Spheniscus*, 6: 19–52.

Sanders, H. 1968. Marine benthic diversity: A comparative study. *The American Naturalist*, 109(925): 243–282.

Scarlato, N. and J. Marcovecchio. 1996. Total mercury in mussels (*Mytilus platensis*) from Mar del Plata, Argentina. In: *Pollution Processes in Coastal Environments*, J. Marcovecchio (ed.), pp. 413–415. Mar del Plata, Argentina.

Schulz-Baldes, M. 1992. Baseline studies on CD, Cu and Pb concentrations in Atlantic neuston organisms. *Marine Biology*, 112: 211–222.

Seelinger, U., L.D. de Lacerda, and S.R. Patchineelam (eds). 1988. *Metals in Coastal Environments of Latin America*, 182pp. Springer-Verlag, Berlin (Germany).

Siciliano, S., L. Lodi, G. Sales, and I. Gonchorosky. 1990. Observacoes de baleias jubarte, *Megaptera novaeangliae*, no Banco dos Abrolhos, costa nordeste do Brasil. *Resúmenes de la Cuarta Reunión de Trabajo de Especialistas en Mamíferos Acuaticos de América del Sur*, p. 40.

Silva, P. and M. Favero. 1998. Kelp gulls (*Larus dominicanus*) and Antarctic limpets (*Nacella concinna*): Their predator–prey relationship at Potter Peninsula and other localities in the South Shetland Islands. *Berichte zur Polarforschung*, 299: 290–296.

Stuardo, J. 1964. Distribución de los moluscos marinos litorales de Latineamérica. *Boletín del Instituto de Biología Marina*, 7: 79–91.

Szidat, L. 1964. La parasitología como ciencia auxiliar de la biogeografía de organismos marinos. *Boletín del Instituto de Biología Marina*, 7: 51–55.

Vaz Ferreira, R. 1982a. *Arctocephalus australis* (Zimmermann). South American fur seal. *Mammals in the Seas*, IV: 497–508. FAO, Rome.

Vaz Ferreira, R. 1982b. *Otaria flavescens* (Shaw), South American sea lion. *FAO Mammals in the Seas*, IV: 477–496. FAO, Rome.

Vaz Ferreira, R. and A. Ponce de León. 1987. South American fur seal, *Arctocephalus australis*, in Uruguay. In: *Status, Biology and Ecology of Fur Seals*, J.P. Croxall, and R.L. Gentry (eds), pp. 29–32. NOAA Technical Report NMFS 51.

Weddell, J. 1825. *The Voyage Towards the South Pole Performed in the Years 1822–1824*, 324pp. Longman Hourst, Rees, Orme, Brown, and Green, London.

Weimerskirch, H., M. Saamolard, F. Sarrazin, and P. Jouventin. 1993. Foranging strategy of wandering albatrosses through the breeding season: A study using satellite telemetry. *The Auk*, 110: 325–342.

Zuquim Antas, P. 1991. Status and conservation of seabirds breeding in Brazilian waters. In: *Seabird Status and Conservation: A Supplement*, J.P. Croxall (ed.), pp. 141–158. ICBP Technical Publication No. 11, Cambridge (UK).

Glossary

aristocracy. Geroniocracy

benthos. Substratum and live organisms related to the bottom of aquatic environments (marine and fresh water).

biomagnification. The process whereby the concentration of a contaminant in tissues increases as the chemical moves from organism to organism through consumption up food chains.

biomarker. A measurable variation or effect in cellular or biochemical processes, structures, or functions induced by foreign substances.

biotop. Plural. Biota a specific support for an ecological community.

biotoxins. Toxic chemicals of biological origin, including the compounds produced during "blooms" of marine microorganisms such as red tides.

bloom. A sharp increase in the density of phytoplankton of benthic algae in a given area.

butyltins. Organic compounds that form complexes with the element tin, including tributyltin, dibutyltin, and monobutyltin. The butyltins are used in anti-fouling marine paints, as catalysts for polyurethane foams and silicones, and as stabilizers for chlorinated polymers.

by-catch. Incidental catch of non-target organisms like dolphins, sea turtles, and seabirds due to fishing operations.

coastal waters. Marine benthic and pelagic ecosystems having substantial influence from the land.

ctenophore. A transparent gelatinous animal floating in surface water at sea, mostly near the seashore. Its common name is "comb jelly".

dinoflagellates. Unicellular planktonic organisms related with the primary production of marine environments. Some species produce the "red tides" that affects humans and many species of marine vertebrates.

dumping. Any deliberate disposal at sea of wastes or other matter, or any deliberate disposal of vessels or other human-made structures.

epizootic. An outbreak of disease in animal populations.

estuary. An ecosystem in which a river or stream meets ocean waters; characterized by intermediate or variable salinity levels and often by high productivity.

eutrophication. Deterioration in quality of aquatic systems due to enrichment with excessive amounts of nutrients, often caused by pollution from human sources; enrichment of a water body with nutrients, resulting in excessive growth of phytoplankton, seaweeds, or vascular plants, and often depletion of oxygen.

exclusive economic zone (EEZ). That part of the marine realm seaward of territorial waters within which nations have exclusive fishing rights.

halieutic. Belonging to the sea resources (*halios* is Greek for "the sea").

halocline. An intermediate layer due to different salinities of water masses.

hypereutrophic. Supreme level of eutrophication.

hypoxia. A state of low oxygen concentration relative to the needs of most aerobic species.

immunosuppression. Reduction in the functional capacity of the immune system.

immunotoxicity. Poisoning of the immune system by foreign substances, resulting in immunosuppression.

intertidal substrate. Surface of coastal area bottoms exposed during low tides.

lipophilic. Substances having a strong affinity for fats; differentially absorbed or solubilized in fats.

mariculture. Controlled cultivation of marine organism in tanks, ponds, cages, rafts, or other structures.

metallothioneins. Low molecular weight proteins that bind certain heavy metals within the cell, rendering them toxicologically inert.

morbillivirus. A group of viruses in the family *Paramyxoviridae* with high virulence that include rinderpest, distemper, measles, and pest-de-petits ruminants and the newly discovered phocine and cetacean morbilliviruses responsible for recent large epizootics of seals and dolphins.

mysticete. Pertaining to the suborder of cetaceans (*Mysticeti*) that is made up of the baleen whales.

nekton. Group of invertebrate and vertebrate species with self-movements, which displace in the water column of aquatic environments.

odontocete. Pertaining to the suborder of cetaceans (*Odontoceti*) that is made up of the toothed whales, including dolphins, porpoises, beaked whales, and sperm whales.

organochlorines. Chemical compounds that include carbon structures complexed with chlorine atoms.

outfall. A place where a sewer, drain, or stream discharges.

P-450 enzymes. Cytochrome P-450 enzymes are involved in the metabolic breakdown and detoxification of organochlorines and other chemical contaminants at the subcellular level.

pinniped. Pertaining to the subgroup (*Pinnipedia*) of the mammalian order *Carnivora* that includes seals, sea-lions, and the walrus.

polychlorinated biphenyls (PCBs). Mixtures of potentially 209 different compounds (referred to as individual PCB congeners) formed by the chlorination of biphenyl with variable numbers of chlorine atoms. These persistent compounds were used in large quantities by industry, and although manufacture has been largely curtailed, they continue to enter the marine ecosystem and are widespread.

radionuclides. Radioactive atoms of elements, which can be environmental contaminants from nuclear weapons testing and processing, nuclear fuels and accidents, and other anthropogenic activities.

red tide. Reddish-brown discolouring of surface water from blooming populations of dinoflagellate phytoplankton; long associated with nutrient pollution, these might be population outbreaks of alien dinoflagellate species.

sirenian. Pertaining to the mammalian order *Sirenia*, modern members of which include the dugong, the recently extinct Steller's sea cow, and three species of manatees.

thalassocracy. Power on the seas realm (*thalassa* → the sea/*kratos* → the power democracy/*demos* → the people).

toxicosis. A pathological condition caused by the action of a poison.

Acronyms

ACCOBAMS	Agreement on the Conservation of the Cetaceans of the Black Sea, Mediterranean, and Contiguous Atlantic Sea
ADCP	acoustic Doppler current profiler
APMW	Asia-Pacific Mussel Watch
ASPACO	Asia-Pacific cooperation project
BC	Brazil Current
BT	butyltin compound
CDV	canine distemper virus
CHL	chlordane compound
COD	chemical oxygen demand
CTD	conductivity, temperature, and density
DBT	dibutyltin
DDT	dichlorodiphenyl trichloroethane
DMV	dolphin morbillivirus
DTH	delayed-type hypersensitivity
EDC	endocrine disrupting compound
EEZ	exclusive economic zone
EMECS	Conference on the Environmental Management of Enclosed Coastal Seas
ENSO	El Niño southern oscillation
EQS	environmental quality standard
ERTC	Environmental Research and Training Centre (Thailand)
FAO	Food and Agriculture Organization
GEOSECS	Geochemical Sectional Studies
GNP	gross national product

GOOS	Global Ocean Observing System
GPS	geographical positioning system
HCB	hexachlorobenzene
HCH	hexachlorocyclohexane
ICSU	International Council of Scientific Unions
ICZM	integrated coastal zone management
IGBP	International Global Biosphere and Geosphere
IMO	International Maritime Organization
IOC	Intergovernmental Oceanographic Commission
IPCC	Intergovernmental Panel on Climate Change
IPG	Iwate Prefectural Government
ISME	International Society for Mangrove Ecosystems
JGOFS	Joint Global Ocean Flux Study
KORDI	Korean Ocean Research and Development Institute
LAC	limits of acceptable change
MBT	monobutyltin
MESSC	Ministry of Education, Science, Sport, and Culture (Japan)
MFC	Malvinas-Falklands Current
MI	metabolic index
MRA	Meteorological Research Association
NGO	non-governmental organization
NIES	National Institute of Environmental Science (Japan)
NOPACCS	North Pacific Carbon Cycle Study
NPO	non-profit organization
OC	organochlorine compound
ORI	Ocean Research Institute, University of Tokyo
PC	Patagonian Current
PCB	polychlorinated biphenyl
PCB	polychlorinated biphenyl
PCD	dibenzodioxin
PCDD	polychlorinated dibenzodioxin
PCDF	polychlorinated dibenzofuran
PDV	phocine distemper virus
POP	persistent organic pollutant
PVC	polyvinyl chloride
SST	sea surface temperature
SWAO	south-western Atlantic Ocean
TBT	tributyltin
TCD	trichlorodiphenyl
TCDD	tetrachlorodibenzodioxin
TCO2	total carbon dioxide
TEQ	toxic equivalent
UNCED	United Nations Conference on Environment and Development
UNCLOS	United Nations Convention on the Law of the Sea
UNEP	United Nations Environment Programme
UNEP-GPA	UNEP Global Programme of Action for the Protection of the Marine Environment from Land-based Activities

UNESCO	United Nations Educational, Scientific, and Cultural Organization
UNU	United Nations University
WCRP	World Climate Research Program
WMO	World Meteorological Organization
WSSD	World Summit on Sustainable Development
WTO	World Trade Organization

Contributors

Zafar Adeel, United Nations University, International Network on Water Environment and Health UNU-INWEH, McMaster University Canada.
Phone: +1-905-525-9140 (ext. 23082)
Fax: +1-905-529-4261
E-mail: adeelz@inweh.unu.edu

Masao Amano, Otsuchi Marine Research Center, Ocean Research Institute, University of Tokyo, Japan
Phone: +81-193-42-5611
Fax: +81-193-42-3715
E-mail: amano@wakame.ori.u-tokyo. ac.jp

Ricardo Bastida, Departamento de Ciencias Marinas Universidad Nacional de Mar del Plata, Argentina, Casilla de Correo 43 (7600) Mar del Plata, Argentina
Phone: +54-223-495-1285
E-mail: biosub@uolsinectis.com.ar

François Doumenge, Oceanographic Museum of Monaco
Phone: +377-9315-3600
Fax: +377-9350-5297
E-mail: musee@easynet.fr

Marco Favero, Departamento de Ciencias Marinas Universidad Nacional de Mar del Plata, Argentina, Funes 3250 (7600) Mar del Plata, Argentina
Phone: +54-223-475-2426
E-mail: marco-patry@hotmail.com

Nobuhiko Handa, Faculty of Information Science and Technology, Aichi-Prefectural University, Japan
Phone: +81-52-851-2191
E-mail: handa@ist.aichi-pu.ac.jp

Zhou Kaiya, Department of Biology, Nanjing Normal University, China
Phone: +86-25-3729111
Fax: +86-25-3783174
E-mail: kyzhounj@jlonline.com

Supawat Kan-Atireklap, Eastern Marine Fisheries Development Center, Department of Fisheries, Thailand

Nobuyuki Miyazaki, Center for International, Ocean Research Institute, University of Tokyo, Japan
Phone: +81-3-5351-6437
Fax: +81-3-5351-6530
E-mail: miyazaki@ori.u-tokyo.ac.jp

Haruhiko Nakata, Department of Environmental Science, Kumamoto University, Japan
Phone: +81-96-342-3380
Fax: +81-96-342-3320
E-mail: nakata@sci.kumamoto-u.ac.jp

Kouichi Ohwada, Faculty of Environmental and Symbiotic Sciences, Prefectural University of Kumamoto, Japan
Phone: +81-96-383-2929
E-mail: ohwada@pu-kumamoto.ac.jp

Tomotoshi Okaichi, 1914–31 Ichinomiya, Takamatsu City, Kagawa, 761-8084 Japan
Phone: +81-87-885-5325
Fax: +81-87-885-5325

Thomas J. O'Shea, US Geological Survey, Midcontinent Ecological Science Center, Colorado, USA.
Phone: +1-303-226-9398
Fax: +1303-226-9230
E-mail: Tom_O'shea@usgs.gov

Bayram Ozturk, Faculty of Fisheries, Istanbul University, Turkey
Phone: +90-216-323-9050
Fax: +90-216-323-9050
E-mail: mmonachus@e-kolay.net

Ayaka Amaha Ozturk, Science and Technology Institute, Istanbul University, Turkey
Phone: +90-216-323-9050
Fax: +90-216-323-9050
E-mail: mmonachus@e-kolay.net

Evgeny A. Petrov, Limnological Institute of the Siberian Division of the Academy of Science of Russia, Russia
E-mail: petrov_evgeny@mail.ru

Diego Rodrìguez, Departamento de Ciencias Marinas, Universidad Nacional de Mar del Plata, Argentina
Phone: +54-223-475-2426
E-mail: dhrodri@mdpedu.ar

Peter S. Ross, Contaminant Sciences Section, Institute of Ocean Sciences, Sidney BC, Canada
Phone: +1-250-363-6806
Fax: +1-250-363-6807
E-mail: rosspe@dfo-mpo.gc.ca

Maricar S. Prudente, Science Education Department, De La Salle University, Philippines
Phone: +63-2-526-5916
Fax: +63-2-526-5915
E-mail: msprudente@yahoo.com

Norberto Scarlato, Departamento de Ciencias Marinas, Universidad Nacional de Mar del Plata, Argentina INIDEP Playa Grande (7600) Mar del Plata, Argentina
Phone: +54-223-486-2586
E-mail: scarlato@inidep.edu.ar

Annamalai Subramanian, Center for Marine Environmental Studies, Ehime University, Japan
Phone: +81-89-927-8194
Fax: +81-89-927-8194
E-mail: subra@agr.ehime-u.ac.jp

Shinsuke Tanabe, Center for Marine Environmental Studies, Ehime University, Japan
Phone: +81-89-927-8171
Fax: +81-89-927-8171
E-mail: shinsuke@agr.ehime-u.ac.jp

Ryo Tatsukawa, 5-911-78 Higasino, Matuyama, Ehime 790-0903, Japan
Phone: +81-89-943-7399

Juha I. Uitto, UNDP-GEF, New York, USA
Phone: +1-212-906-5724
E-mail: juha.uitto@undp.org

Machiko Yamada, Aqua Research Center, Kitakyushu City Institute of Environmental Science, Japan
Phone: +81-93-882-0333
Fax: +81-93-871-2535
E-mail: machiko_yamada01@mail2.citv.kitakyushu.jp

Index